The Biology of Agricultural Systems

The Biology of Agricultural Systems

C. R. W. SPEDDING

Deputy Director and Head of Ecology Department,
The Grassland Research Institute, Hurley

and

Professor of Agricultural Systems,
Department of Agriculture and Horticulture,
Reading University

1975

ACADEMIC PRESS
London New York San Francisco
A Subsidiary of Harcourt Brace Jovanovich, Publishers

ACADEMIC PRESS INC. (LONDON) LTD.
24/28 Oval Road,
London NW1

United States Edition published by
ACADEMIC PRESS INC.
111 Fifth Avenue
New York, New York 10003

Second printing 1977

Library of Congress Catalog Card Number: 74-18505
ISBN: 0-12-656550-3

Printed in Great Britain at The Spottiswoode Ballantyne Press
by William Clowes & Sons Limited, London, Colchester and Beccles

Preface

This book is primarily concerned with the underlying biology of agricultural systems and with the role of animals and plants within them. It is not greatly concerned with the biology of the components of agriculture, however, and it does not deal with particular animal or crop species and how they are farmed. It is much more concerned with the way in which biological processes are integrated into whole agricultural systems, with their multiplicity of purposes, products and resources.

The aim has been to relate the underlying biological processes to the agricultural systems of which they are parts (or sub-systems, as I have argued in this book). Man's dependence on agriculture as applied biology is so great that some knowledge of its purposes and functions could be said to be indispensable to a broad education. It is argued that this kind of agricultural biology should form part of a general education, serving as a link between the natural and the social sciences.

Although the book is intended for students of agriculture or of applied biology, therefore, I hope that it might be useful in a much wider context. Because it deals with a very broad view of the biology of agriculture, it cannot afford to assume that the reader has specialized knowledge of zoology, botany or of any of the other relevant disciplines. I have tried to keep the language simple and, although specific biological systems are used to illustrate principles, it is the latter that I have tried to define, describe and convey.

October, 1974 C. R. W. S.

Acknowledgements

I wish to acknowledge my indebtedness to Professor E. K. Woodford, the Director of the Grassland Research Institute, and to the Governing Body, for facilities made available to me in writing this book and for permission to use both published and unpublished data.

I am most grateful to the many colleagues who have allowed me to draw freely upon their results and who have helped in the compilation of figures and tables. These contributions are acknowledged in the text but I wish especially to record my thanks to Miss A. M. Hoxey and Miss J. M. Walsingham, both of whom helped me enormously during the preparation of the book. Thanks are also due to Mr. N. R. Brockington, Mr. R. V. Large, Dr. J. G. W. Jones, Professor E. K. Woodford and Professor E. H. Roberts for helpful discussions and for reading critically some parts of the manuscript.

I wish to acknowledge permission to reproduce figures, tables and photographs as follows:

the editor of *Agricultural Progress* (Tables 1.1 and 1.2); the editor of *Outlook on Agriculture* (Table 1.3 and Figs 2.1, 2.4 and 2.5); Dr. Helen Newton Turner (Table 1.3); Sydney University Press and the editor of the Proceedings of the 3rd World Conference on Animal Production (Table 2.1 and Figs 3.5 and 3.6); the British Steel Corporation (Fig. 2.16); Dr. D. H. Lloyd and The Organisation and Methods Society (Figs 2.21 and 2.22): the editor of the Proceedings of the 2nd World Congress on Animal Feeding (Figs 2.23 and 2.24); Dr. S. Housley and the editor of *Laboratory Practice* (Table 3.1); the British Farmer and Stockbreeder Ltd. (Fig. 3.9); Dr. J. E. Storry and the Society of Dairy Technology (Fig. 3.11); Professor Harold J. Evans and Plenum Publishing Co. Ltd. (Fig. 3.15); Dr. F. W. T. Penning de Vries and Academic Press (Fig. 3.16); Professor J. L. Harper, P. H. Lovell and K. G. Moore and Annual Reviews Inc. (Figs 4.1 and 4.3); Dr. M. Gadgil and the University of Chicago Press (Fig. 4.2); Dr. D. B. Arkcoll and Blackwell Scientific Publications Ltd. (Fig. 4.5); Dr. J. S. Olson and Associated Book Publishers Ltd. (Table 4.16); Dr. Magnus Pyke and Heinemann Educational Books Ltd. (Table 4.17); Dr. D. E. Reichle and Springer-Verlag (Table 4.20); Bailliere, Tindall and Cassell (Fig. 5.3); Reinhold Publishing Co. (Fig. 5.6); Cambridge University Press (Table 5.32, Figs 5.8–5.12 and 10.2); R. V. Large (Figs 5.8–5.10, 5.12 and 10.2);

Professor H. A. Regier (Table 5.32); Professor A. N. Duckham, G. B. Masefield and Chatto and Windus (Fig. 8.3); Dr. J. D. MacArthur and The Clarendon Press for Figure 8.4, from "Farming Systems in the Tropics" by Hans Ruthenberg; and Professor Howard J. Dittmer for his ready agreement to my use of his material in constructing Tables 4.1–4.7.

I should also like to record my thanks to Mrs H. L. B. Stone, Miss J. Springall and Miss W. A. Haynes for their quick and accurate typing, and to the Grassland Research Institute for permission to use Plates I and II.

Finally, I wish to record my appreciation of the helpfulness of the publishers during the production of the book.

Contents

To:
G. R. S.

1 The Purposes of Agriculture

It seems so obvious that the main purpose of agriculture is to produce food, that further discussion appears to be superfluous. A moment's thought results in immediate qualification, to include other products as well as food, but this is a relatively minor matter. The important point is that the whole idea is a serious oversimplification.

To obtain an immediate insight into the problem, consider the facts that world agriculture could easily produce vastly more food, that over-production has been a major problem of many regions, that considerable efforts and incentives have been devoted to reducing production in many areas, and that a great many people are hungry. Of course, there are difficulties of distribution, of wealth with which to buy food and of the food itself, since it is not always produced where it is most wanted.

Nevertheless, it has to be recognized that, just like any other industry, agriculture is practised for a wide variety of reasons, to make money being one of the most common.

The diversity of purposes for which agriculture is carried out in the poor and populous regions of the world has been discussed by Bunting (1971). His list can be expanded into the general purposes given in Tables 1.1(a) and 1.1(b), based

TABLE 1.1 The major purposes of agriculture
(a) The provision of a product

Main products	Purposes included
Human food of plant origin of animal origin	Feeding local population Export and substitution for imports
Animal feed of plant origin of animal origin	Feeding local farm animals Export for farm animals or pets
Raw materials for Industry of plant origin of animal origin	Processing and manufacture of clothing and furnishings Industrial food production
Recreational facility	Farm zoos and other amenity provision
Money	Profit Return on investment

(b) The use of a resource

Main resources	Purposes included (other than production)
Land	Preservation of amenity
Labour	Provision of work Provision of a way of life
Money	Investment (incl. attraction of foreign exchange)
Physical inputs	Use of locally produced inputs Use of inputs imported for other reasons

mainly on the production of products and the use of resources, as summarized by Spedding (1973). A glance at Table 1.2 (Turner, 1971) shows why it is necessary to include other aspects of animal performance than those concerned strictly with products. Virtually all the purposes of agriculture, however, can be included in the phrase "provision of a product or service and the use of a resource". What makes the operation an agricultural one, therefore, cannot simply be its purpose, for many of the purposes listed could be found in clearly non-agricultural industry. It is, in any case, probably better to recognize that there are many marginal activities that are difficult to distinguish clearly from agriculture. Common examples are shooting, fishing and hunting for sport, the keeping of animals in zoological parks, industrial production of microbial protein (see Mateles and Tannenbaum, 1968) and the spinning of synthetic meat substitute.

Equally, it is sometimes difficult to be sure when an operation like battery egg-production, or many of the indoor horticultural production systems, have become sufficiently divorced from normal agriculture that they should no longer be included.

In most definitions, agriculture involves deliberate use of land for cultivation of crops or the feeding of animals and it is perhaps still most helpful to regard the use of the land for these purposes as most characteristic of agriculture. Whether a particular animal or plant is regarded as "agricultural" depends chiefly on whether it gives rise to useful products or services, but some element of control of production is usually regarded as characteristic also.

Where a purpose can be related to a main product or resource, however, it does not mean that other products and resources are unimportant. Quite often, by-products are of great total value and resources other than the main one may easily be of very great importance because of their scarcity or cost. In many farming systems several products are involved and in virtually all systems the total balance of product values and resource costs is what determines whether they can be operated profitably.

This raises a very important question. Even where profit is not the major purpose of an agricultural operation, it is unlikely that anyone can continue to operate it unless it makes a profit. Loss-making operations can be imagined but

TABLE 1.2 Types of domestic animals and their contributions to man's material needs (from Turner, 1971)

Animal type	Contribution
Buffalo	Traction, transport, meat, milk, leather, horn
Camels:	
Alpaca	Textile fibre, meat
Camel	Traction, transport, meat, milk, fats, textile and other fibre, leather, fuel, fertilizer (bone and dung)
Llama	Transport, textile fibre, meat
Vicuna	Textile fibre
Cat	Hunting (of rodents)
Cattle	Traction, transport, meat, milk, fats, leather, textile and other fibre, fertilizer (bone and dung) fuel, horn, glue, blood (for food and for serum in biological products)
Dog	Transport, meat, hunting (for food), guarding (man or animals), working (cattle or sheep), guiding (the blind), tracking
Elephant	Traction, transport, ivory
Equines:	
Ass	Traction, transport, meat, milk, leather, fuel
Horse	Traction, transport, meat, leather, fibre (hair from mane and tail), fuel, fertilizer, blood (for serum)
Mule	Traction, transport, fertilizer
Ferret	Hunting (for food)
Fur-bearers:	
Chinchilla, ermine, fox, marten, mink and sable	Fur
Goat	Meat, milk, textile and other fibre, leather, fertilizer, horn, blood (for serum)
Monkey	Harvesting (coconuts)
Pig	Meat, leather, bristle, fats
Reindeer	Traction, transport, milk, meat, leather, textile and other fibre
Rabbit	Meat, textile fibre, fur, laboratory animal (biological products)
Rodents:	
Guinea-pig	Meat, textile fibre, laboratory animal (for preparation of biological products)
Hamster	Laboratory animal (biological products), fur
Mouse	Laboratory animal (biological products)
Rat	Laboratory animal (biological products)
Sheep	Textile and other fibres, meat, milk, fats, leather, fur, fuel, fertilizer, horn, transport
Yak	Transport, textile fibre, meat, milk, leather

where such losses are deliberately absorbed it is really because the agricultural enterprise is only a part of a larger system. The latter might be an individual's way of life or a large commercial organization; alternatively, it is possible to argue that a very high value is in fact being placed on carrying out the enterprise (or even its physical presence) and that this is not entering into the accounts.

The main practical point is that however efficiently an agricultural system produces its products or uses its resources, calculated in non-monetary terms, it will not be operated for long unless it is financially viable. One of the difficulties in discussing purposes is that products and resources may be interchangeable (i.e. one man's product is another man's resource) and they therefore vary with the nature and scale of the agricultural system considered. It may be useful to outline briefly the more important products and resources used in farming.

Products

Since a farmer is involved in growing crops or keeping animals, his products can be most readily classified into those derived from one or the other.

Animal Products

These may be:

(a) Livestock for breeding or further growth, performance or production;
(b) Parts of the slaughtered animal; or
(c) Secretions, excretions, etc. of the live animal.

The range is thus very large. (a) Any agricultural animal can be sold as a product from one farmer to another at any stage: perhaps the most common would be fertile eggs and young animals for rearing, older animals for "fattening" or "finishing", and mature animals for breeding, lactation, egg production or wool production. (b) Virtually all parts of all relevant animals have a use but by far the most important is represented by the lean meat of the carcase. (c) The most important products from live animals, other than the live animals themselves, are milk from female cattle, sheep, goats, buffalo, etc., eggs from female birds (chiefly), wool and hair that is shorn or plucked off the animal at intervals, and excreta for use as fertilizer or fuel.

In addition to these obvious products, however, there are less well-known items, such as semen from male animals (one of the few specifically male products), and a very important category of services. The most important of the latter are related to power, for traction or transport. Historically (from surprisingly early times–Murray, 1970), male animals surplus to breeding requirements have been castrated, mainly to make them more docile and therefore more manageable, and such castrates have been used for meat and wool production and for both traction and transport.

Some animal products are used to feed other animals, but this tends to be an inefficient process and most animals kept for food production are herbivorous or omnivorous. Presumably there was no incentive to domesticate animals that competed for the same kind of food as man himself required and man has developed by displacing carnivores rather than by eating them. There are other reasons why carnivores were domesticated as pets, originally living on waste products of the human diet. Carnivores are limited in numbers by the large herbivore populations needed to support them, in turn limited by plant production per unit area of land. Man's omnivorous habits have allowed a much greater human population, therefore, than would have been possible if he had simply displaced the major carnivores. These developments are illustrated in Figs 1.1 and 1.2.

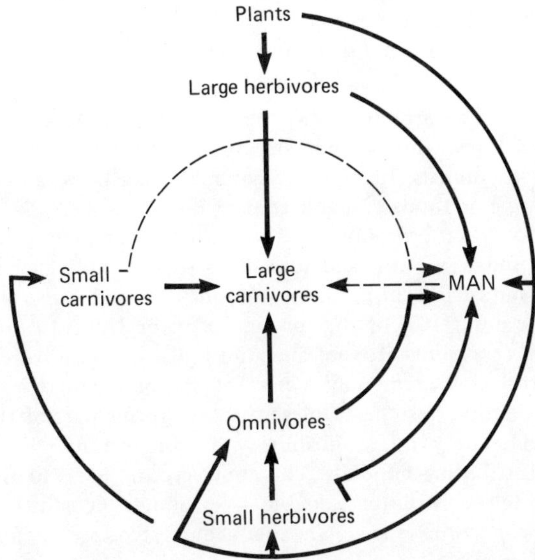

FIGURE 1.1 The large carnivore as the ultimate consumer in the food chain.

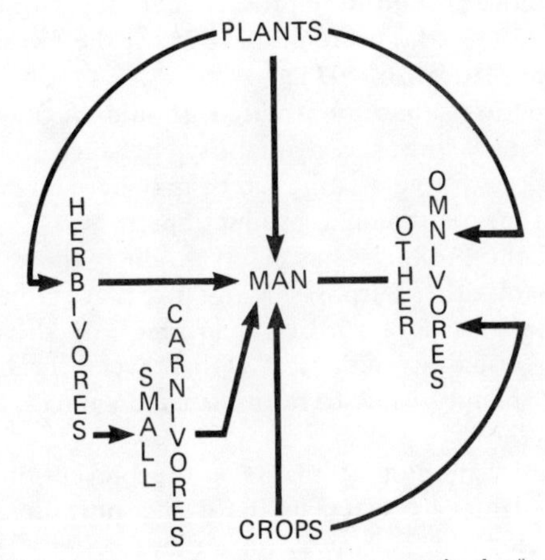

FIGURE 1.2 Man as the ultimate consumer in the food chain.

Crop Products

Since even carnivorous animals ultimately depend upon crop production to sustain their herbivorous prey, plant growth has been the essential basis of agriculture. In the future, the industrial extreme of the agricultural range may alter this situation but it seems unlikely that the basic importance of crops will be challenged in the foreseeable future.

Crop Products are of three main kinds:

(a) Underground storage organs,
(b) Above-ground storage organs and
(c) Above-ground stems and foliage.

Any of these raw products may be subjected to further processing and extraction procedures, whether they are intended for direct human consumption or for feeding to animals. Indeed, such animals may be regarded as one of the possible processing methods available to us.

Of course, not all crop products are used for food and enormous quantities of such crops as timber, cotton and jute are grown for use in construction and manufacturing industries: an immense amount of wood is also still used for fuel (Schery, 1972); about 10% of that produced in the U.S.A., *ca.* 30% in Europe, *ca.* 75% in Asia (excluding Russia) and up to 90% in South America and Africa.

Since the crop may form only a part of the plant, there are other outputs from crop production systems. Unless the root forms part of the product it is rarely used deliberately but contributes to "soil fertility" by adding organic matter and redistributing minerals. The above-ground parts of the plant, on the other hand, often contribute valuable by-products in addition to the main product. The proportion of the plant that is harvested (see Chapter 6 for further discussion) varies greatly: it tends to be high in annuals and much less in perennials, of course. Thus an annual apple crop forms only a small part of the dry weight of the (above-ground) tree producing it, but with the wheat crop 35% of the dry matter, 72% of the protein, and 52% of the digestible energy may be contained in the grain (Hodgson, 1971).

It is therefore natural that by-products should be more important from short-lived crop plants: the sheer quantity of such by-products and the proportion that can be utilized as animal feed may have an important bearing on the profitability and future of animal production.

In addition to all the products that plants produce, they also are used—just as with animals—for many other purposes. Many plants are grown for their amenity value to man, sometimes as a sort of "clothing" of the soil, as with grass, sometimes for their visual attraction, as with flowers, shrubs and trees. Others are grown for their amenity value to both man and animals, as with shelter belts and shade trees.

Another important function of plants is the prevention or control of soil erosion. Grassland plants are often used for this purpose, on steep slopes to prevent water carrying the soil away in gulleys or on flat land to prevent wind-erosion. Sand dunes are often stabilized by planting with salt-resistant species such as marram grass (Hubbard, 1954). Of course, it is not obvious exactly where the line should be drawn that defines what is or is not a strictly agricultural purpose but, in the examples given in the previous paragraph, the primary purpose is often linked to maintaining an agricultural usage of the land.

Recreational Facilities

It will be clear from the foregoing that both plants and animals may be used for amenity purposes, whether combined with agricultural purposes or not.

It can be argued that such purposes have little to do with agriculture, but crops and animals require management and the scale of this need in the future might be very large indeed. This is only one of the reasons why agriculturalists should contribute to the debate on amenity use of land, for the cheapest and most effective management may be found by integration of land use for different purposes. This discussion will be further pursued in Chapters 8 and 9, but it may be helpful to illustrate the topic at this point with an example.

One of the biological facts that illustrates the need for management is the way in which bare soil does not remain so for long. In fact, bare soil has little to recommend it for most amenity purposes but what grows by itself if the soil is left alone may be little better. Grassland is probably the most useful single ground-cover for amenity use, but it has to be defoliated, partly because it is more generally useful when short and partly because, in the U.K. at least, it progresses to scrub or woodland if it is not defoliated.

In some cases, especially if the land is flat, grassland can be frequently cut, as with lawns and cricket pitches, and may even be better so. But on more difficult land, with steep slopes, rocky outcrops, stones and boulders, or boggy patches, cutting may be impracticable and on a large scale it may be grossly uneconomic.

The cheapest way to maintain grassland in a short state would be to graze it with large hervibores. Several factors have to be considered in the choice of animal: ability to survive the climate; ability to thrive in the physical environment; ability to protect itself against predators, people and dogs; relative lack of need for attention, to save labour; lack of danger to other animals and people; and grazing habits that are non-destructive to the environment.

Agricultural herbivores, such as cattle and sheep, have most of the qualities required and suggestions have been made (Spedding, 1974) that some of the least domesticated species might be best: e.g. the Soay sheep (Plate I, facing p. 198) does not require shearing and is very hardy. In addition, of course, there are distinct possibilities of obtaining from agricultural animals some products that might in total be of significant value. At the very least, they would help pay for their management.

It might be supposed that agricultural animals would require much more management than would be the case with wild species. It is not conducive to amenity, however, to allow populations of large herbivores to manage themselves. This can only lead to the presence of dead and sick animals and may also lead, at times, to emaciated individuals and disease epidemics. If the number of animals is sufficient to keep the grassland short at the height of the herbage growing season, it will require management, in terms of culling and population control, and some supplementary feeding at times of shortage.

There is much to be said, therefore, for the use of farm animals that are accustomed to management and for the employment of agriculturalists who are skilled and experienced in the task. On the other hand, there is everything to be said for exploring the possible use of non-agricultural animals, such as red deer (Bannerman and Blaxter, 1969) for this purpose or for integrated agricultural and amenity use of land.

Money

Some agricultural enterprises are undertaken solely in order to make money, although not necessarily as profit. Very few farmers can afford to ignore financial considerations, although probably few put money-making (in the sense of maximizing profit, for example) in the front of their minds to the exclusion of all other purposes.

The importance of making money in agriculture is so great that no proper consideration of agricultural systems can leave it out. It is an essential integral component of these systems and cannot be adequately dealt with by attaching costs and prices to essentially biological systems. Money is involved in the use of resources like labour, for example, that may not even be included in a consideration of biological systems.

The impact of this on agricultural biology, is chiefly twofold. First, it establishes the framework of agricultural systems (see Chapter 2) within which biological systems operate, and, secondly, it means that biological optima are not necessarily synonymous with economic optima. Sometimes the issue is further obscured by the fact that some satisfactions (such as the sight of Hereford cattle from the window) are not valued amongst the outputs, and some costs (such as nervous strain on the manager) are not included amongst the monetary inputs. Apart from this, however, it must not be expected that the price of a product will necessarily reflect in any way the biological cost of producing it, or that the costs of inputs will necessarily reflect their biological value or the costs of producing *them*.

Resources

Money is a very important resource and, without it, little use of other resources may be possible. The most characteristic resources of agriculture are land, labour, management, animals and plants, and, especially of modern intensive farming, fertilizer. There are many other important inputs, such as fencing materials and water, that vary greatly from situation to another.

Land

Originally, this could be regarded as the indispensable resource: it consisted of soil in which to anchor and nourish crop plants and, perhaps, a source of water.

Increased intensity of agricultural use of land tends to diminish the role of the soil somewhat, and it is regarded more as a medium to which nutrients are added, as required, for crop production. It is not easy to assess the importance of the myriads of living creatures in the soil, in terms of the productivity and stability of agricultural systems, but it would be unwise to ignore these aspects (see Chapter 8).

The structure and nature of the soil nevertheless have a very important bearing on what is agriculturally possible, although it is difficult to classify soils in terms of agricultural potential. The topography of the land and the climate to which it is exposed may be linked together as another powerful determinant of potential.

Labour

The labour of the farmer may be augmented by that of general or specialized assistance, permanent or casual, but in highly developed agricultural systems the need for labour is greatly reduced by the use of machinery. The latter requires considerable capital investment and is a major consumer of energy (see Chapter 7). Whether this progressive reduction in the number of people directly involved in agriculture is a good thing or not may be debated. It has certainly led to the development of systems of agriculture in the developed western world that are inappropriate or irrelevant to the needs of many developing countries.

Animals and Plants

Since these can only be produced by other animals and plants, they must be represented by the products already considered. Breeding stock, new-born animals, eggs, semen, partly-reared animals and milk products for animal feeding are the more important animal resources. Seed, tubers, cuttings and other propagative parts are the most important plant resources for crop production. In addition, there are all the crop products that are produced for feeding to animals and those that form the basis of industrial agricultural processes.

Fertilizer

Plant nutrients may be obtained from decomposing plant roots (a fairly continuous process), from the decomposition of ploughed-in vegetable matter (green crops or stubbles), from the faeces and urine of grazing animals, from the manure (animal faeces and urine mixed with bedding*) of stabled animals, or from artificial fertilizers. The last are now used in enormous quantities, being chiefly compounds of nitrogen, phosphate, potassium and calcium. Modern, intensive agriculture could not achieve anywhere near its present levels of production without these inputs and they often represent a significant cost in an enterprise.

* In its modern, intensive form this may have no bedding but occur mixed with water, as "slurry".

The Purposes of Agricultural Systems

Agricultural systems are generally constructed and operated to fulfil one or more of the purposes described in this chapter. Their success cannot be judged, therefore, except in relation to the particular purposes for which they were designed or carried out. Nevertheless, each system may be carried out for different purposes at different times, in different places and by different people. In discussing agricultural systems (see Chapter 2) and the ways in which they may be studied, described and understood, these points should be borne in mind. The immediate purpose of a system may be obvious; for example, to produce beef, but the purposes of producing beef (to eat, for money or to employ people) may have an over-riding influence on the details of the system.

Man's Basic Needs

In spite of the fact that the purposes of agriculture are so varied, it is worth attempting to describe what man needs and hopes to obtain by his agricultural activity. The more sophisticated of modern requirements can only be generalized about: "amenity" is a useful term to describe one category of purposes but it may mean, in detail, as many different things as there are people.

In fact, this is now true of most of the purposes mentioned in this chapter and the major exception is food. As man has become less dependent on animals for work, less dependent on crops for shelter materials, and less dependent on either

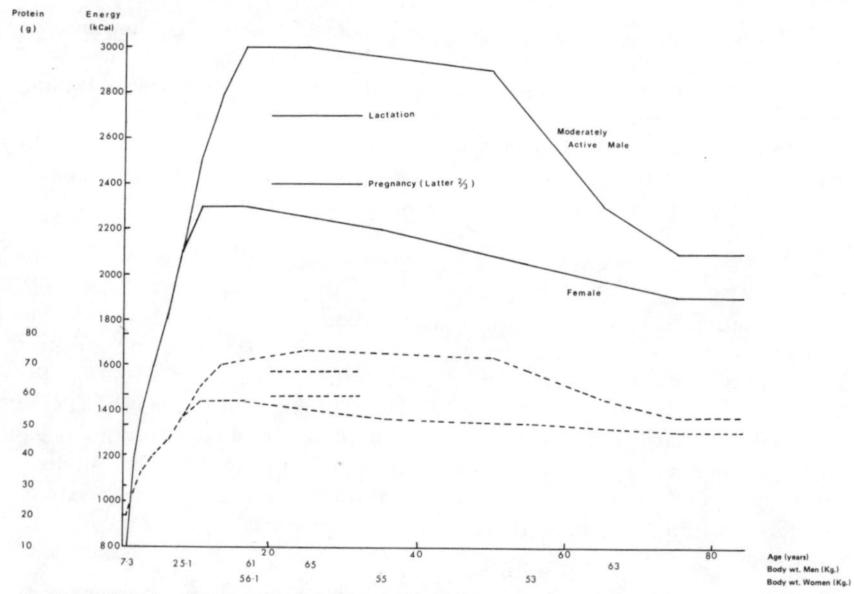

FIGURE 1.3 Recommended daily intakes of energy and protein in the U.K. (based on Rep. Publ. Health & Med. Subj., 1970). Energy (kcal) ———; Protein (g) - - - - -.

for clothing, so his needs for these items become less basic. Thus the demand for fur is no longer basic at all; it is related to the satisfaction of quite sophisticated desires and therefore takes many forms.

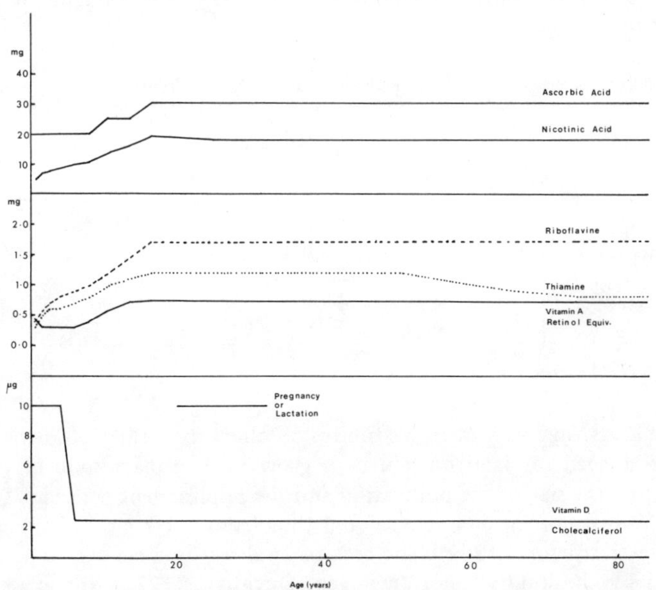

FIGURE 1.4 Recommended daily intakes of vitamins. All lines except the one for pregnancy relate to males (based on Anon., 1970).

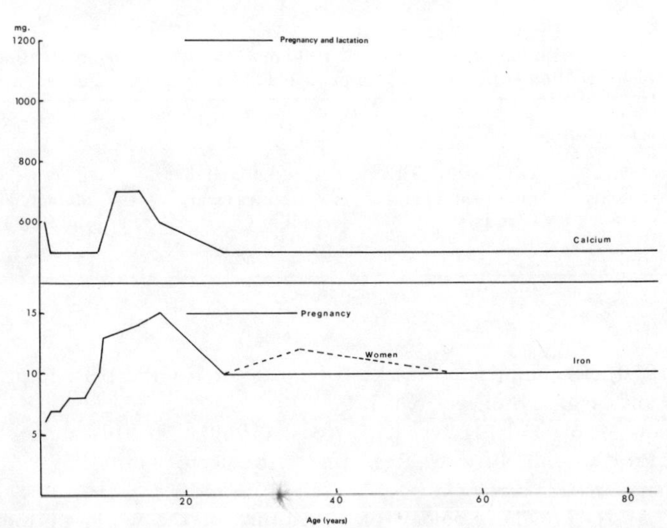

FIGURE 1.5 Recommended daily intakes of minerals (based on Rep. Publ. Health & Med. Subj., 1970).

For food, however, there is still a basic need for energy, protein, minerals and vitamins. This is illustrated in Figs 1.3 to 1.5, for U.K. conditions. Such needs vary, not only with body weight but also with activity and climatic conditions. Table 1.3 shows the variation in protein and energy requirements in different regions of the world.

TABLE 1.3 Regional variation in protein and energy requirements per caput per day

Region	Energy (kcal)	Protein (g)
North America	2710	74
Western Europe	2660	74
Central America	2310	62
Central Africa	2220	60
India	2200	60
Pakistan	2280	53

Source: F.A.O. (1970)

Since the vast majority of man's food is obtained from the land, there may be said to be a need for land on which to grow it. The magnitude of this need depends upon the size of the population and the requirement per man. The latter varies with environment and activity and thus from one region of the world to another (see Carpenter, 1969). The area of land required per man is also greatly influenced by the kind of diet envisaged. Duckham (1971), for example, has calculated a range of more than four-fold, depending on the nature of the diet alone (Table 1.4).

TABLE 1.4 Area required (in acres) to feed 1000 moderately working men of body weight 65 kg (mean) at 1969 U.K. standards for protein and energy. (After Duckham, 1971)

Diet			
Very austere with no animal proteins or fats	Less austere with milk protein and with fats	Good mixed diet with meat and milk	An actual diet, commercially produced 1962–65
238	470	783	1530

References

Anon. (1970). Recommended intakes of nutrients for the U.K. Rept. Public Health and Medical Subjects, No. 120.

Bannerman, M. M. and Blaxter, K. L. (eds) (1969). The Husbanding of Red Deer. Proc. Conf. Rowett Res. Inst. Aberdeen, Jan. 1969. Aberdeen University Press.

Bunting, A. H. (1971). Ecology of agriculture in the world of today and tomorrow. Presented to a Symposium at the National Academy of Sciences, Washington, D.C.

Carpenter, K. J. (1969). *In* "Population and Food Supply" (Sir Joseph Hutchinson, ed.), Chapter 5. Oxford University Press.

Duckham, A. N. (1971). *In* "Systems Analysis in Agricultural Management" (J. B. Dent and J. R. Anderson, eds), Chapter 16. John Wiley and Sons, Australia.

F.A.O. (1970). Provisional Indicative World Plan for Agricultural Development. 2.

Hodgson, R. E. (1971). *J. Dairy Sci.* **54** (3), 442-447.

Hubbard, C. E. (1954). "Grasses: a guide to their structure, identification, uses and distribution in the British Isles," p. 263, Penguin, Harmondsworth.

Mateles, R. I. and Tannenbaum, S. R. (eds) (1968). "Single Cell Protein". The M.I.T. Press.

Murray, J. (1970). "The First European Agriculture". Edinburgh University Press.

Schery, R. W. (1972). "Plants for Man" (2nd edition). Prentice-Hall, Englewood Cliffs, N. J.

Spedding, C. R. W. (1973). *Agric. Prog.* **48**, 48–58.

Spedding, C. R. W. (1974). *Biologist* **21** (2), 62–67.

Turner, Helen Newton (1971). *Outlook on Agriculture* **6** (6), 254-260.

2 Agricultural Systems

Definitions

It is necessary to have strict definitions, but they are often rather arid and tend to sacrifice helpfulness to precision. The most important of the definitions relevant to this book are given in the Glossary (p. 242). In the text, however, the object of definitions is to be helpful, even if the terminology is thereby used less strictly.

In this sense, then, **systems** are simply sets of components that interact with each other in such a way that each set behaves as a whole entity. **Agricultural systems** are simply those which have an agricultural purpose: they are also **ecosystems** if, as is commonly the case, they have one or more living components (see Dale, 1970). Since nothing has been said about size of components or systems, it follows that the latter could be very large (e.g. embracing a considerable agricultural area of the world) or very small (e.g. a single crop plant) and the components could vary from single cells or organs up to whole flocks, herds or crops.

If such variety seems impossible to handle, consider the range within animals that zoology covers, or the range of plant forms, sizes and functions that is coped with by botany. The answer in all these cases is to classify the individual objects into useful groups.

Classification of Agricultural Systems

There are several reasons why it is important to classify agricultural systems. The first is that the number of different kinds of system with which anyone can deal is quite small: it is not possible to cope, even in discussion, with the many thousands of individual systems that currently exist, quite apart from considering new ones.

Thus, unless we can generalize about categories such as milk-producing systems or arable farming systems, we shall not be able to discuss them usefully, or plan them, carry out research into them, legislate for them or even name them. This need to classify individuals into groups is essential in all subjects and carries with it certain dangers. The fact is that there are nearly always several different ways of classifying objects and it is essential to choose the most useful for any particular purpose. The dangers relate to the misuse of classification schemes, generally by using them for quite different purposes from those originally intended.

It must be obvious, however, that we will want to use a classification of agricultural systems for several different purposes. When looking at milk-producing systems, for example, it is useful to distinguish between those based on the cow and those based on the goat or the buffalo. Nor does the argument end there, of course: so a whole scheme of classification is required. Whatever may be stated, as a result of research or experience, about one individual system, it is vital to know to which other individual systems it applies. We wish to be in a position to say that it applies to all others of the same class and, in order to do this, we have to know to which class it belongs and what are the characteristics of the class. So, for different purposes, it may be necessary to have different classification schemes but, within any one of them, it should be possible to use the different classification levels (e.g. orders, families etc.) for somewhat different purposes.

A simple example may be helpful at this point. Suppose something is discovered about a cow: the usefulness of the information will depend a great deal on knowing whether it applies to cows, mammals, herbivores, ruminants, large animals, female animals, milk-producers or only, say, to high-yielding milk-producers, one breed of cow or only to the particular cow used. Yet the same cow could be a representative sample of any of these groups; so it is necessary to be sure what we are using the individual cow for and to recognize that we are rarely interested in particular cows themselves (see Chapter 10).

The construction of even one complete classification scheme for agricultural

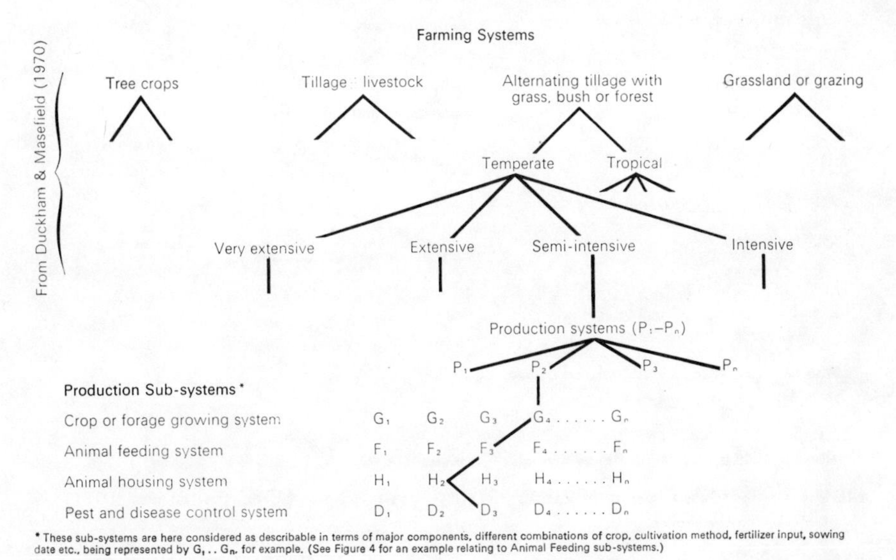

FIGURE 2.1 Classification of agricultural systems, greatly abbreviated (from Spedding, 1971).

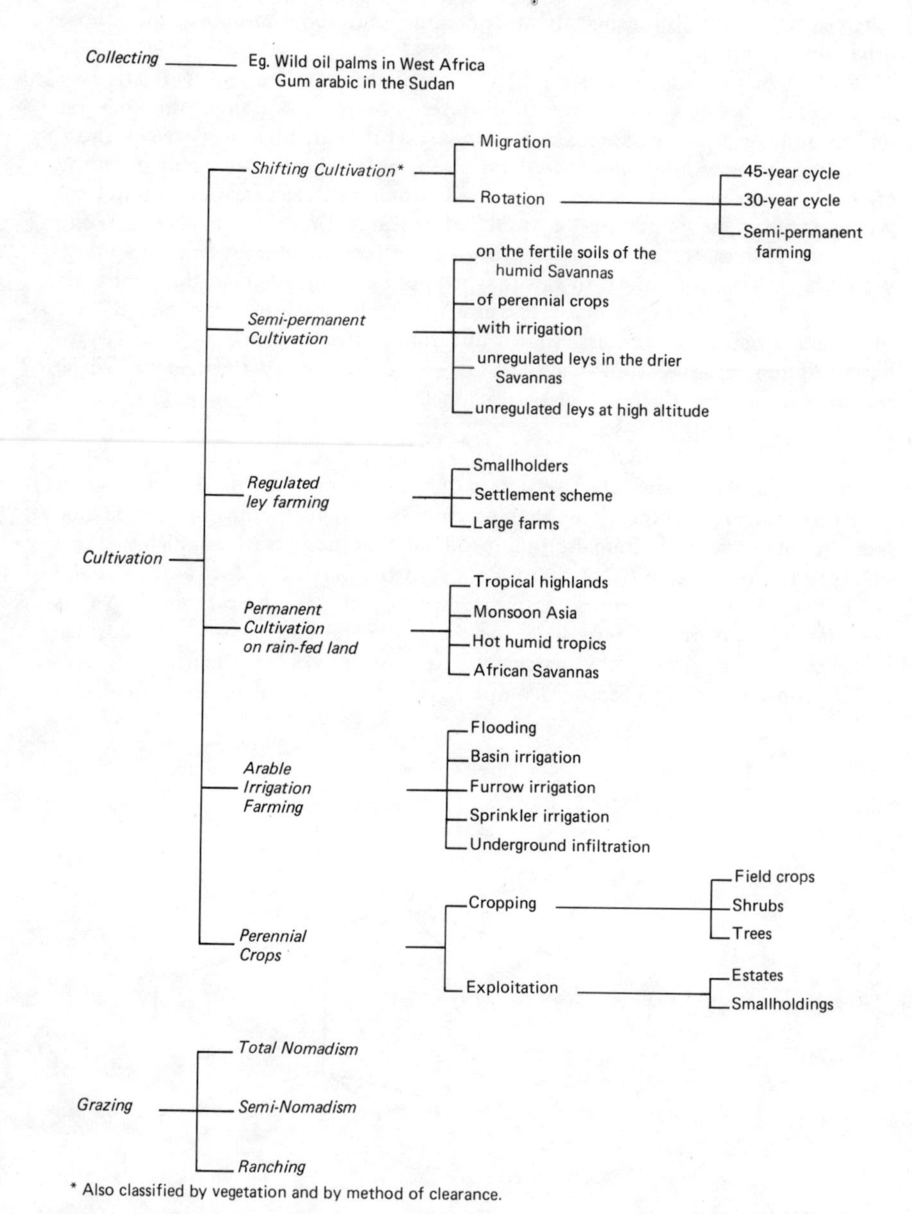

FIGURE 2.2 Classification of farming systems in the tropics (after Ruthenberg, 1971).

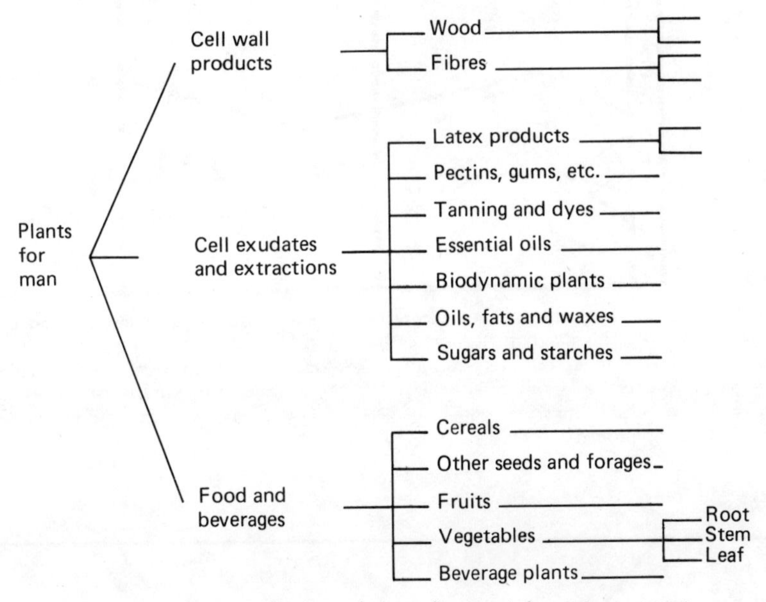

FIGURE 2.3 Classification of plants for man (after Schery, 1972).

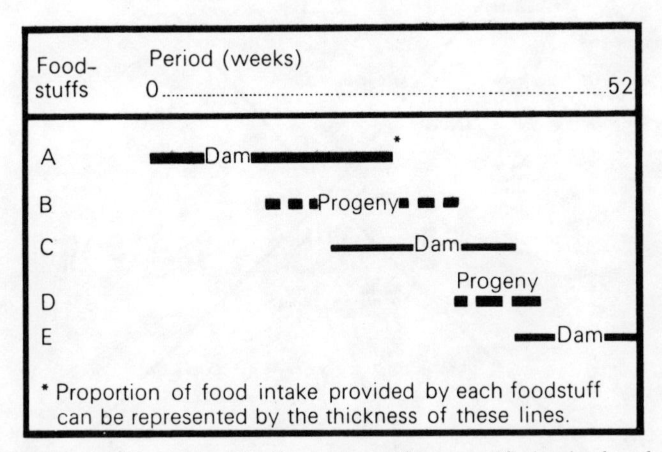

FIGURE 2.4 Classification of feeding systems (for a specified animal and product), described in terms of the foodstuff used, the animal receiving it and the period and season of feeding (weeks 0–52 representing a calendar year) (from Spedding, 1971).

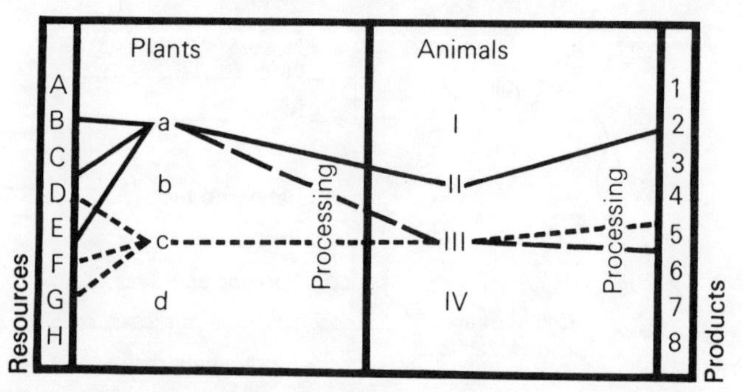

FIGURE 2.5 Diagrammatic representation of a scheme for the classification of biological systems. In this diagram, Resources (A–H) refer to plant nutrients, Products (1–8) refer to crop or animal outputs, and Processes refer to the conversion of resources to products via plants (a–d) and/or animals I–IV: the number and kind of processes can be increased as required. lines (————, – – – – –, - - - - - -) represent different biological systems (from Spedding, 1971).

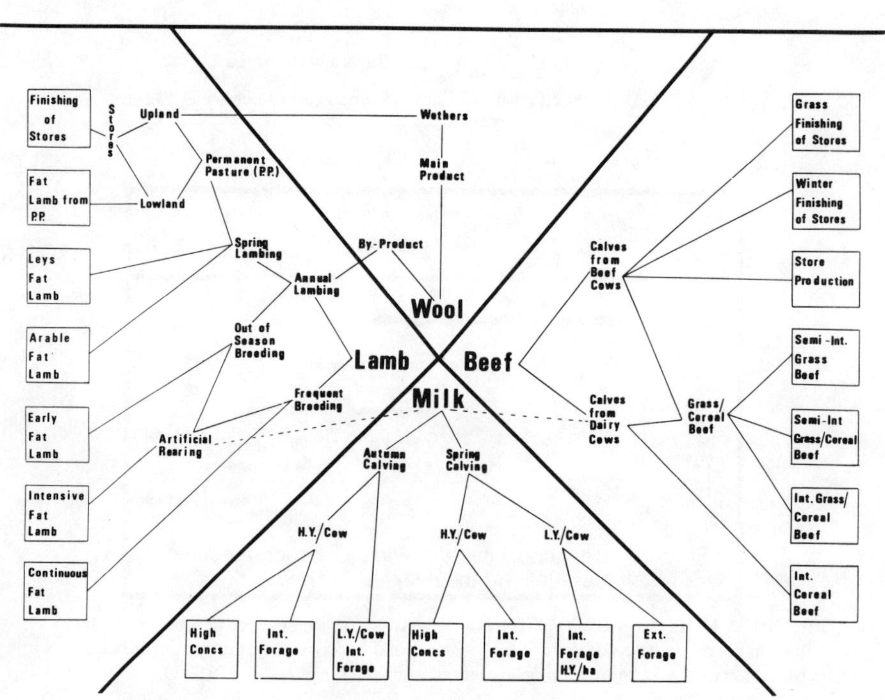

Notes: Int. = Intensive; Ext. = Extensive; Concs. = Concentrate Feeds;
H.Y. & L.Y. = High and Low Yielding.

FIGURE 2.6 A classification of ruminant systems (from Spedding, 1972).

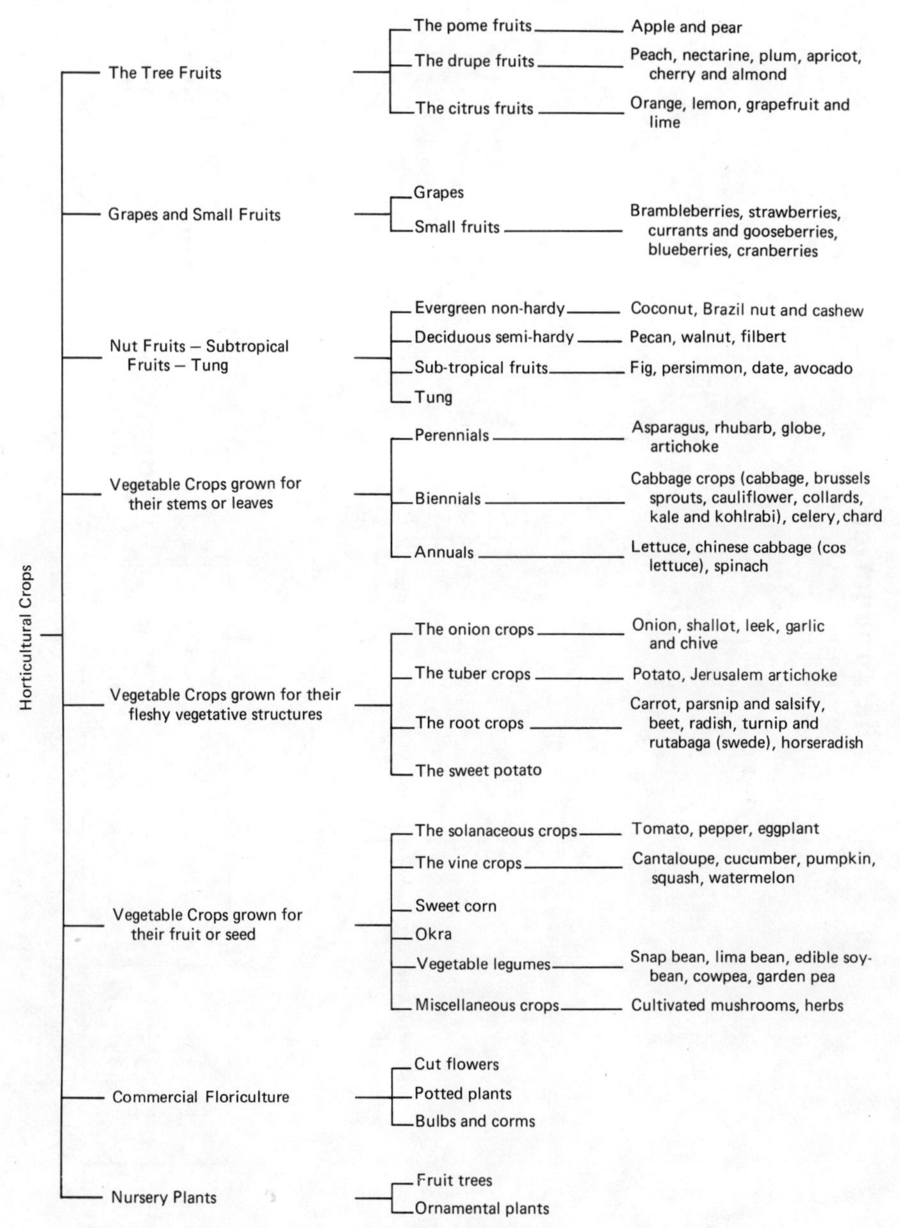

FIGURE 2.7 Horticultural classification (after Edmond *et al.*, 1957).

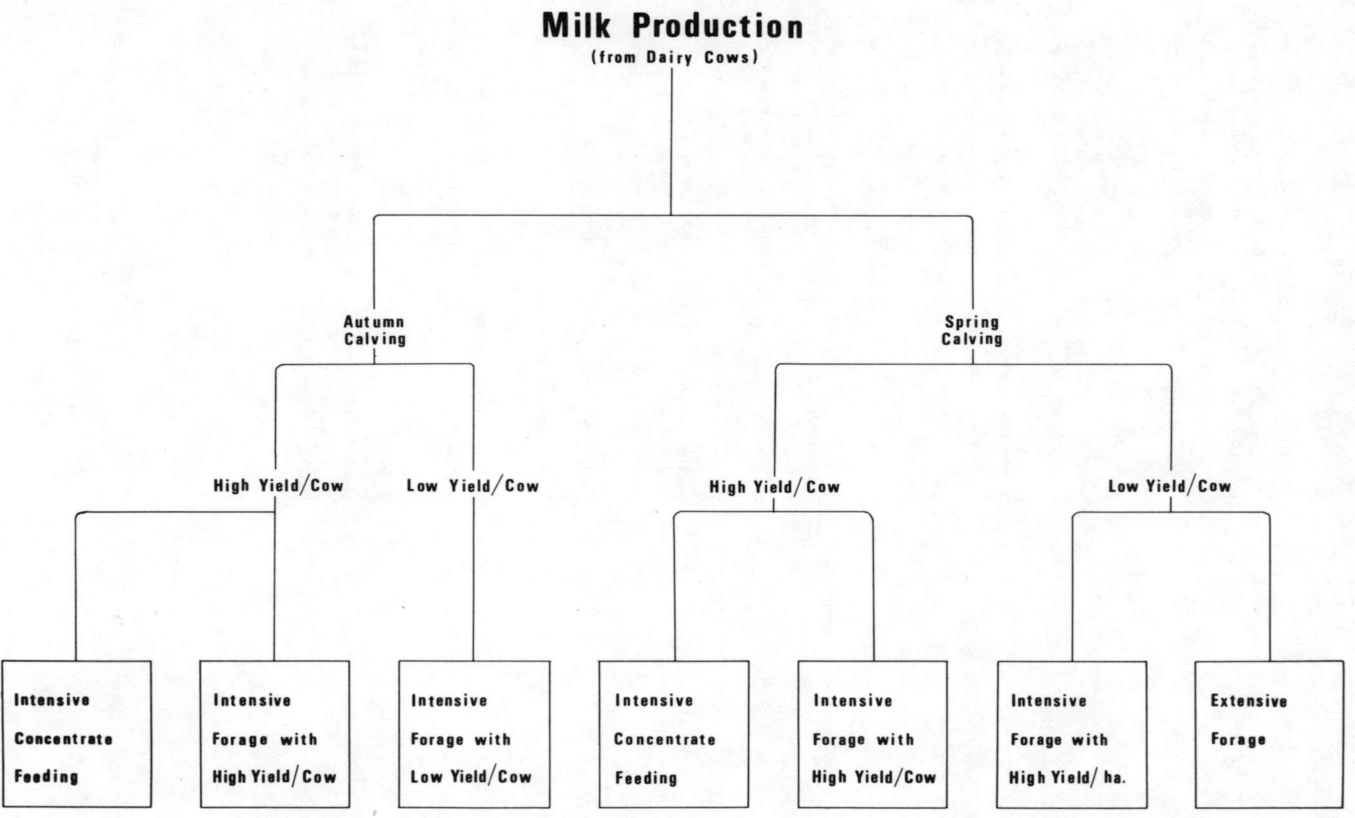

FIGURE 2.8 A classification of milk production systems involving dairy cows.

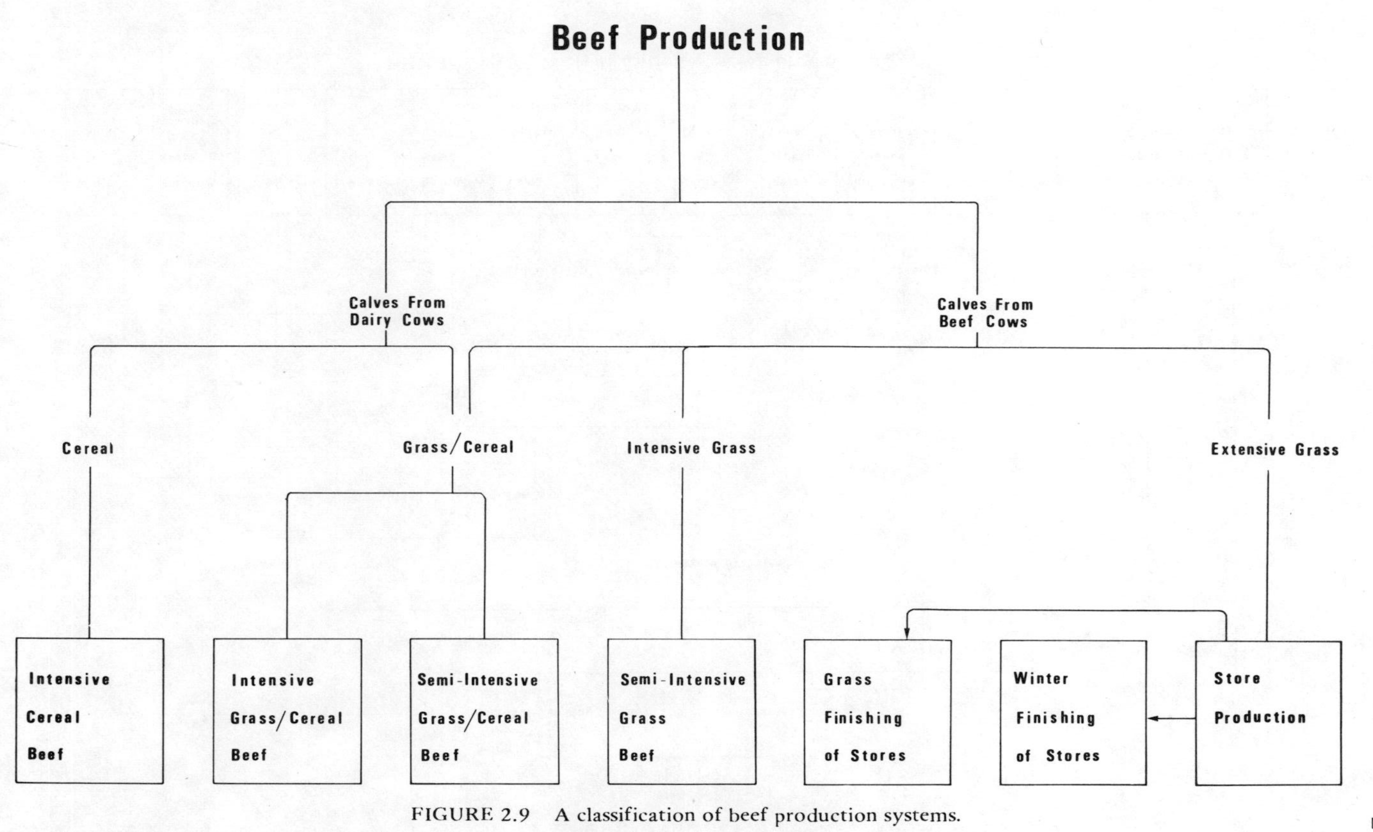

FIGURE 2.9 A classification of beef production systems.

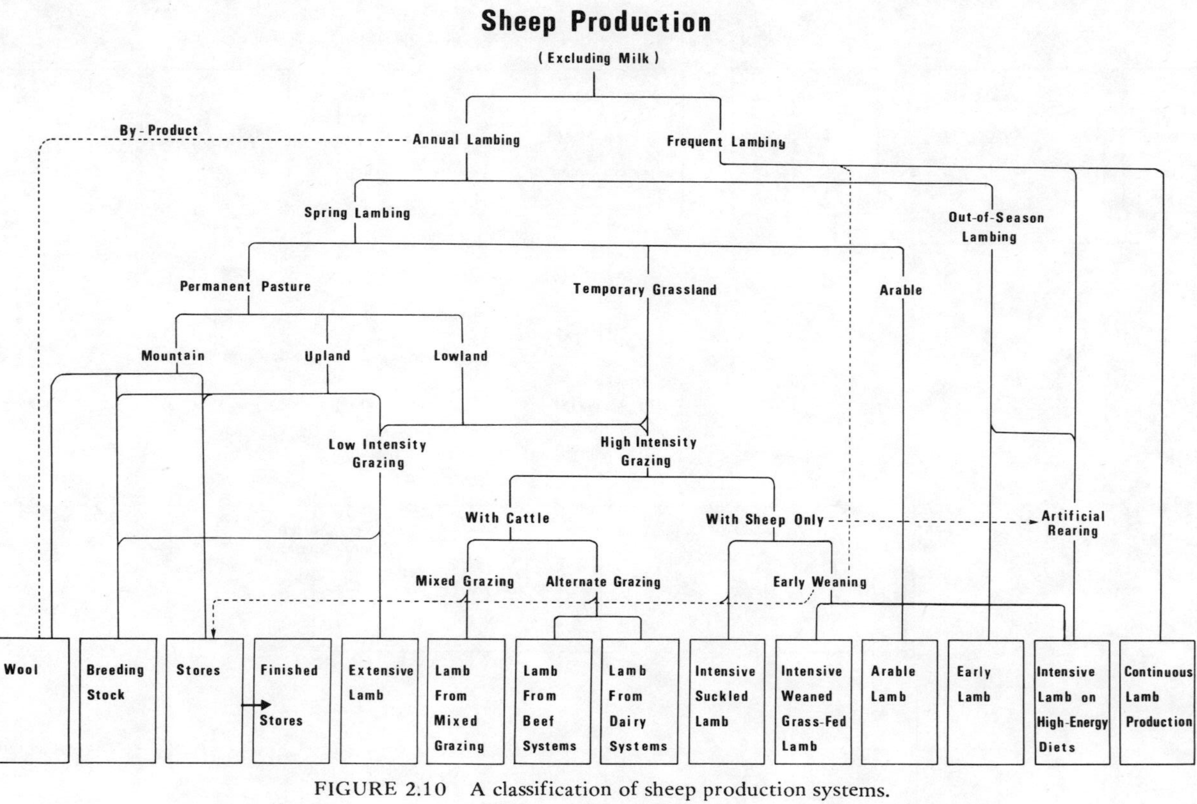

FIGURE 2.10 A classification of sheep production systems.

systems is a very large task and very little progress has yet been made. For the purposes of discussing agricultural biology, however, it is only necessary to consider the most important ways in which such systems might be classified and to be aware of the broad framework that would be most useful. Some of the important ways of classifying are related to the purposes for which agriculture is carried out.

For example, systems that give rise to the same main product or use the same main resource can be readily grouped. Within product-classes, it is then of interest to consider different ways of producing the same product, whether these involve different plants or animals or environments; within any one of these groups there are then sub-divisions characterized by more detailed descriptions of the precise ways in which resources are converted to products.

Products appear a good starting point, since production is a major agricultural purpose, but, since profit is generally required, the choice of system is initially based on some consideration of the available resources as well. Indeed, in the sense that climate and physical environment are the basic resources to be used, it might be expected that the broadest classification of agricultural systems would be related to them.

Several schemes are shown in outline in Figs 2.1 to 2.10, in order to illustrate the levels of classification and the way in which actual named systems can fit into a world-wide scheme.

Many other schemes have been employed and will continue to be employed for different purposes. The high cost of labour in industrialized countries has meant that agriculture has had to become labour intensive. Because of this importance of labour, it is sometimes sensible to classify systems on the basis of the number of "Standard Man Days" involved, combined with "farm type" (Garner, 1972).

The Naming of Systems

It is not merely convenient that recognized systems should have names, it is an essential shorthand. It is helpful if the name is short and related to the purpose, content or structure of the system. Where a classification scheme exists, the name of a system immediately tells you how it is related to other systems and thus a good deal about it, because it will be at some point in a hierarchy of systems that helps to characterize it. In zoological terms, for example, the name spiny anteater (*Tachyglossus aculeata*) immediately indicates at least two features of importance, but its place in the class Mammalia and the order Monotremata also tells us that it is a mammal, and therefore possesses true hair and feeds its young on milk, and that it is an egg-laying marsupial, because these are attributes of all members of that class and order. None of this necessarily tells you what a spiny anteater looks like or how it lives and functions and there is no one picture that can possibly do all this.

So it is with agricultural systems: naming them is one thing and describing them is quite another, but the name becomes an essential code in finding any description that exists. Before systems can be usefully described, however, it is necessary to form a picture of what they are like, what they consist of and what they do.

Visualizing Agricultural Systems

The first step in understanding an agricultural system is to picture it in your mind; the second is to describe it. These are similar activities but the latter involves so much more detail that different methods are required. Now, although a mental picture *may* contain a great deal of detail, it does not follow that it is all essential or that all the essential bits are present.

At a very crude level there is no problem: indeed, our ability to discuss an animal, such as a sheep, depends upon each of us being able to visualize one that is sufficiently similar to that pictured in the minds of the others. These crude images can be manipulated up to a point but beyond this they are no longer adequate. Because these things are common knowledge, each of our sheep will have four legs, one head and so on, and, if we wish, we can specify colour, how many lambs it is suckling or how much wool it grows.

But, as we quantify the image, and add more complicated relationships, it becomes impossible to hold it all in the mind unaided. For many agricultural systems, this is so from the outset. So, even visualizing the system initially requires some assistance, generally by what is appropriately called a "visual aid". It might be thought that, because there are so many different agricultural systems, the problems of "visualizing" them might be uniquely different. The situation appears more hopeful than that, however, and some generalizations are possible about the ways in which systems may be pictured. In general terms, what we need is a "concept" of systems, a general picture that is basic to all or to a large group of systems: this process is sometimes called "conceptualization".

The trouble with most of our mental pictures is that they are rather vague, yet highly specific, i.e. we have rough impressions of particular things. Conceptualization is almost the opposite, concepts consisting of clear and precise pictures of generalizations.

Conceptualization

Any concept of an agricultural system must include the following, as a minimum:

(a) Purpose—the purpose for which the system is being carried out;
(b) Boundary—a means of deciding what is inside and what is outside the system;
(c) Context—the external environment in which the system operates;
(d) Components—the main constituents that are related to form the system;
(e) Interactions—the relationships between components;
(f) Resources—components lying within the system that are used in its functioning;

(g) Inputs—used by the system but emanating from outside it;
(h) Products or Performance—main, desired outputs;
(i) By-products—useful but incidental outputs.

Possibly other features should be included; too little is yet known to be dogmatic about this but the problem is to reduce the picture to its essentials.

If this list is approximately right, then the simplest picture of agricultural systems in general will contain these and only these items. In a sense, the list is itself a picture, but it is not a very good one because it does not show (visually) the way in which items are related to each other. Figure 2.11 shows one simple way of visualizing the agricultural systems concept but there is room for a great deal of improvement in the layout of the diagram. (This is true of most such diagrams and it is part of their role to make possible a parallel evolution of the picture in the mind and that on the paper.) In order to limit the complexity only a few resources and components have been shown and only a few factors indicated for some of the probable interactions. In most agricultural systems, the complexity is vastly greater than this and quite different diagrams are needed.

A very useful kind of diagram has proved to be one based on concentric circles (Spedding, 1972a; 1973b): these allow the major output of the system to be shown centrally (and thus given visual prominence and importance) and the factors influencing that output to be arranged in concentric circles such that where possible, importance and distance from the centre are positively related.

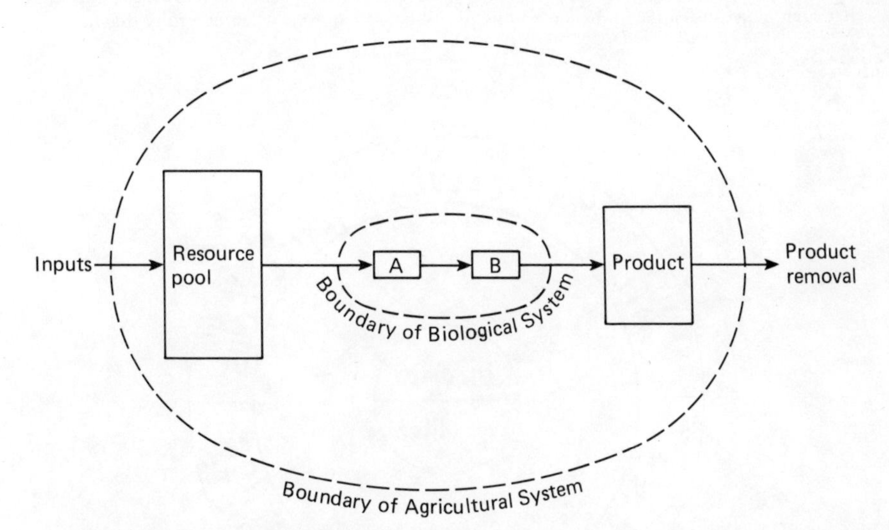

FIGURE 2.11a The simplest possible representation of the way in which a biological system (with only two components, A and B) may operate within an agricultural system converting resources into products. For example, nitrogenous fertilizer might be one input, adding to the soil nitrogen pool, used by grass (A), subsequently eaten by cows (B) producing milk.

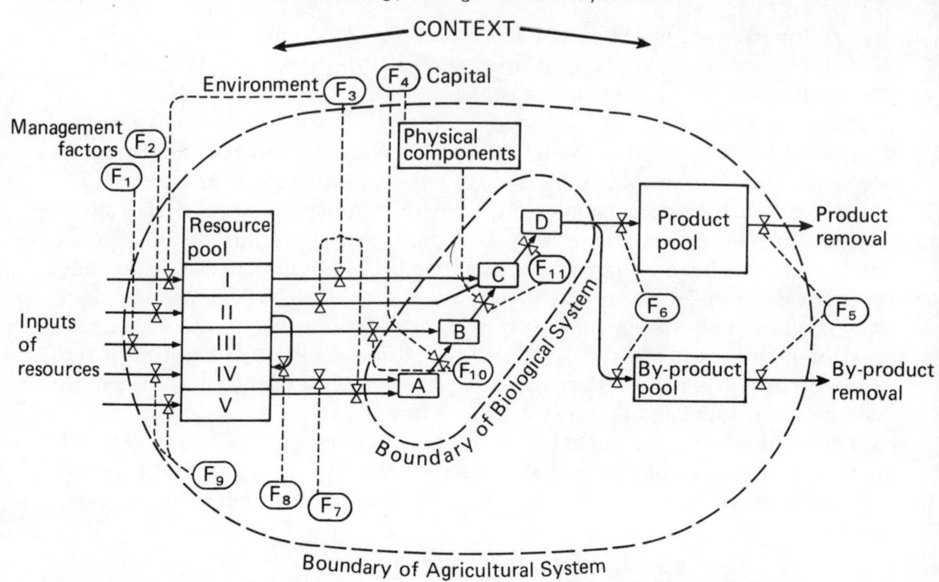

FIGURE 2.11b An expansion of Fig. 2.11a, to include more biological components and to recognize that several resources are usually involved. Biological components (A, B, C and D) are shown as a simple system. All continuous lines with arrows represent flows of material and each rate of flow is determined by factors ($F_1 - F_n$) linked by dotted lines to the valve-symbols (⋈). Only some of the factors and interactions are shown and it is clear that such diagrams can soon become unhelpfully complex, even when carefully drawn.

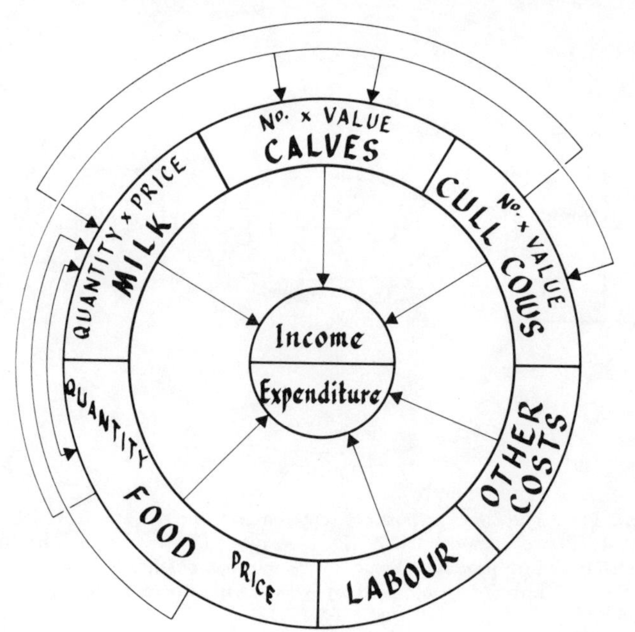

FIGURE 2.12 Profit from milk production (per hectare).

This means that if the whole picture is too large and complicated, it is always possible to reduce it by dealing only with relatively inner circles, with some confidence that the more important factors are retained and the less important omitted. It is also possible to indicate interactions by circular and radial lines only and thus minimize the confusion that results from a mass of criss-crossing arrows.

As it happens, most agricultural systems have a ratio as their major output (see Fig. 2.12), since we are rarely interested in the *absolute* output of a product: generally, the interest is in the output of product per unit of resource. In most cases, there is more than one product and a whole range of resources and *for agricultural systems* the situation frequently has to be expressed in monetary terms, the central ratio representing profit or return on investment or something of this kind. (The situation is generally different when considering the underlying *biological systems*—see Chapter 3.)

The general approach is illustrated in Figs 2.13-2.16, showing how the picture can be systematically built up (in practice, it is helpful to superimpose detail, especially of interactions, by means of a transparent sheet).

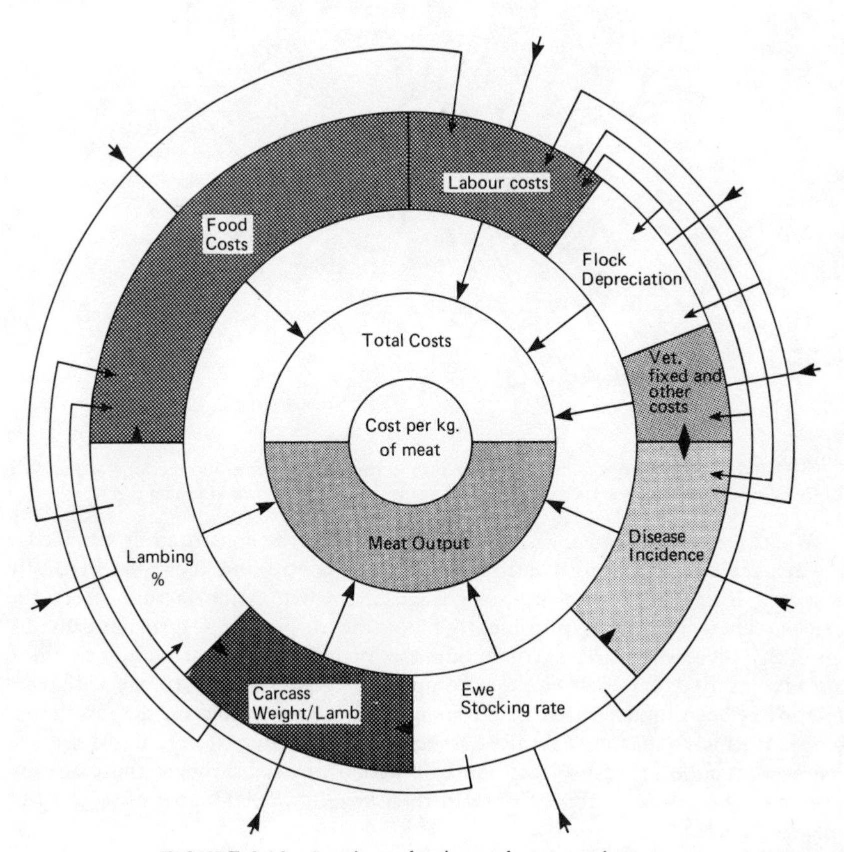

FIGURE 2.13 Lamb production and costs per hectare.

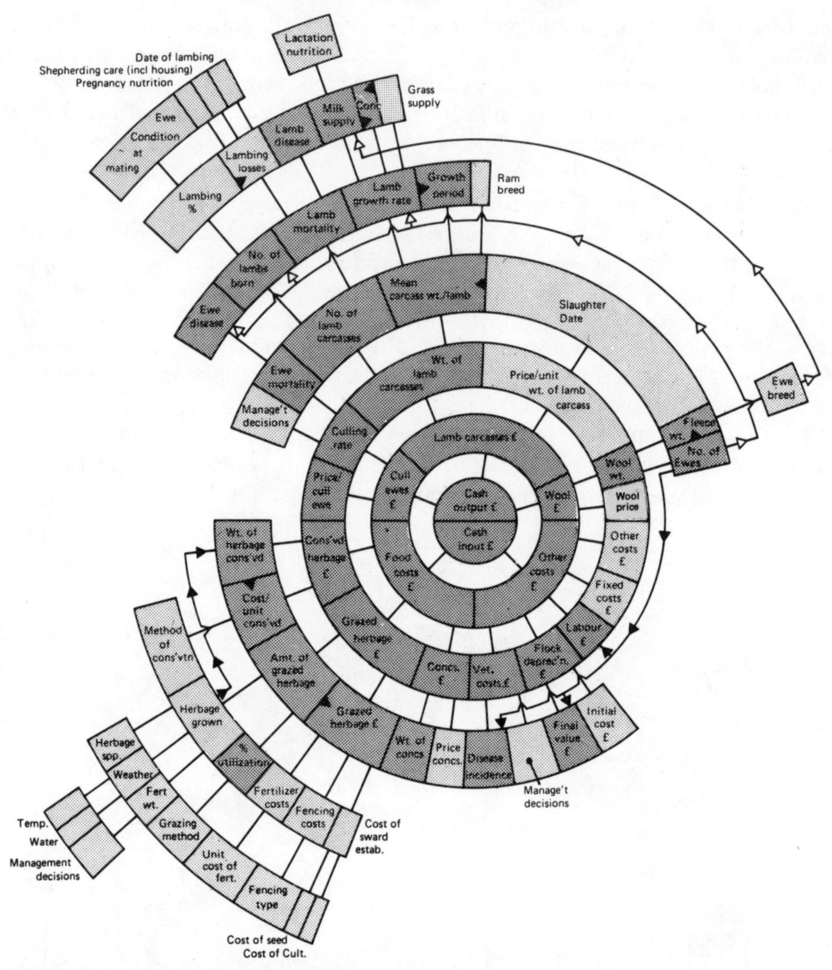

FIGURE 2.14 Components in sheep production (only some interactions are shown. Both Figs 2.14 and 2.16 are shaded to show a sub-system related to stocking rate).

Whatever the system, and almost independent of the constructor's knowledge of the subject, it is possible to agree on the centre and then systematically enquire, from those who are knowledgeable, what major factors affect the centre. Thereafter, it is possible to build the diagram up systematically by inserting successive circles containing the main factors that influence those already inserted. As each circle is completed, it is necessary to check that no factor has been omitted that could *directly* affect something on the next inner circle. If this condition is satisfied, then all other (outer circle) factors are less important, if only because they must be part of, or act through, those already inserted. The relevant interactions can then be determined before proceeding to the next circle.

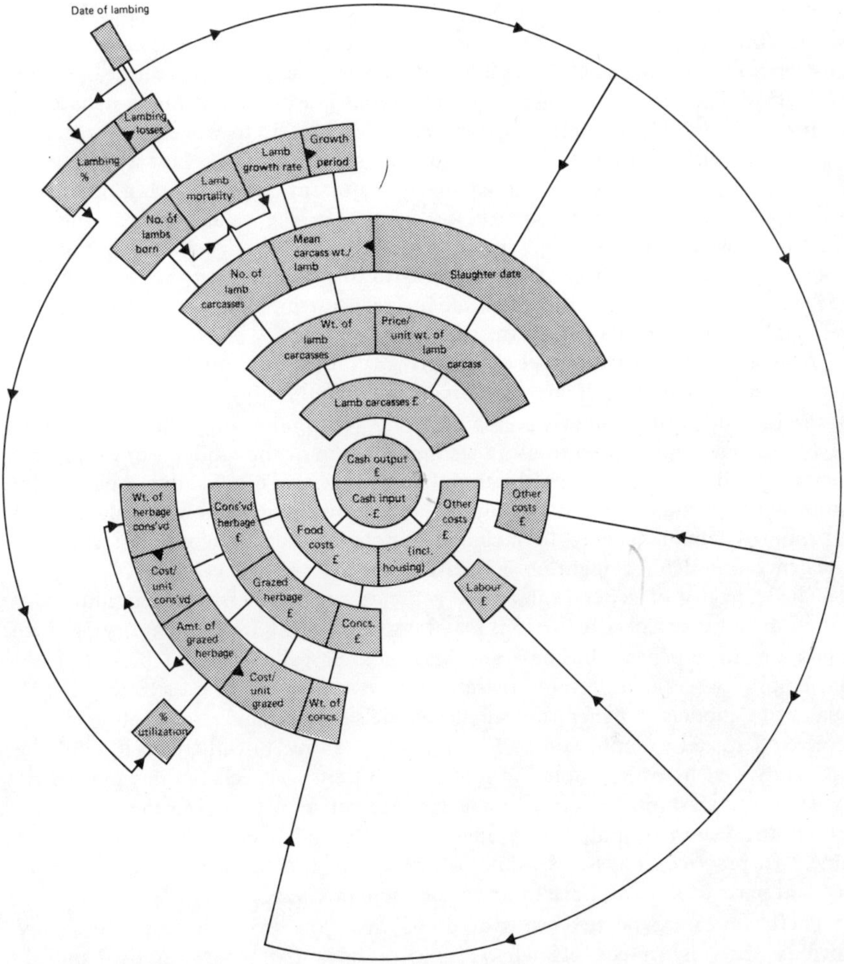

FIGURE 2.15 Effect of date of lambing (a sub-system of Fig. 2.14).

A more complete and realistic example is given in Fig. 2.16 and the problem of even reproducing such diagrams emphasizes the need to take seriously the way they are constructed. If they do not clarify complex situations, they have failed: superficially, however, they appear very complicated. The fact is, of course, that agricultural systems *are* complicated and the complexity does not disappear because we choose to ignore it. Some further examples are given in the Appendix, and the same approach can be applied to all agricultural systems. The crucial test is whether the method helps in visualizing an agricultural system, its content, purposes and boundaries and whether it provides an unambiguous basis for discussion leading to improvement of the picture. If not, or in any case, we have to ask, what is the alternative?

These diagrams represent qualitative models of agricultural systems and it is worth considering whether the constituent relationships can be quantified. The answer is clearly that this is possible but that it would take a great deal of time and effort. On an even larger scale, the same questions are being debated in relation to "world" models (Meadows *et al.*, 1972), as to whether they can be quantified with sufficient accuracy to be useful and how best to use their predictive properties. It is accepted that a major function of such models is to indicate the probable consequences of changes in the values and assumptions used in constructing them and that manipulations of this kind assist in thinking about very complicated situations. Predictions only apply to the real world to the extent that the model includes all the relevant information and adequately reflects the future behaviour of the real world.

A general qualitative model of an agricultural production process includes a great many possible systems and prediction would only apply to particular versions. Specifying one version would involve quantifying the relationships involved, but it might also result in a simplification of the model. For example, a general model of animal production has to have within it the capacity for animals to reproduce in any month of the year but, in general, it is unlikely that all months would need to be included. Similarly, housing would be present in the general model but might not actually occur in a specific version.

This capacity of general models to generate specific versions is an important function but even specific versions may be so complicated that quantifying them could be an expensive business and testing their validity might be extremely difficult. It is very important, therefore, to consider the role of large, general, qualitative models in generating smaller models that can be more easily handled.

Even a casual examination of Figs 2.12-2.16 shows how difficult it would be to study even one whole agricultural system, especially in a classical experimental fashion by varying one component at a time and measuring the effect on system output, on an adequate scale and over a sensible period of time.* In practice, it is not feasible to proceed in this fashion (see also Chapter 10) and parts of systems have to be studied separately.

There appear to be two ways of doing this. The first, and often the only obvious way, is to cut off what seems to be a part that can be handled. Obviously, this is done in as sensible a manner as is possible, but it is often, nonetheless, a rather arbitrary procedure. This is sometimes rather like wishing to understand an animal (as a living system), finding it too complicated and difficult to study the whole animal and chopping off a leg at a time for separate study. Not only are there obvious doubts about the usefulness of going about it in this way, there are serious difficulties in ever combining the information so gained in such a way that understanding of how the whole animal functioned is improved (Spedding, 1973a). This is mostly due to the fact that important interactions between parts, studied separately, would never be studied at all.

The second approach is to try and identify *sub-systems.*

* See Spedding and Brockington (1974) for a discussion of the problems of experimentation with agricultural systems.

Agricultural Sub-Systems

Before considering the question of sub-systems and whether they can be extracted from the circular models already described, it is as well to emphasize that any kind of diagram can be well—or badly—constructed and that the use that can be made of a diagram depends upon sound construction (and even on the quality of the drawing).

The way in which these circular diagrams are built up has been outlined briefly but there are many alternative ways of arranging factors and of deciding on the best way of grouping and naming them. Some of this has to be left to individual judgement but experience suggests that some rules are emerging that can be summarized as a developing methodology: details of the latter are given in the Appendix.

The sub-system approach differs substantially from the arbitrary consideration of parts of systems. First, it must be recognized that every component of a system can also be regarded as a system on its own and be examined in the same way, including having a model built to describe its essentials. Such a model, however, would be at a different level appropriate to the system being examined. Thus a model of a herd of cows would have individual cows as constituents; a model of mixed grazing would have both cows and sheep. The addition of sheep, however, is an extension of the model at the same level, that of the individual animal. But a model can also be constructed of the individual cow and the behaviour of the model would be intended to reflect that of the cow. In this case, the individual components might be organs and this is regarded as being at a different level. Most models have to be limited to three levels: (i) that of the component; (ii) that of the system itself and (iii) that of the larger system of which the system modelled is a component.

Secondly, within a system and at the same level, there are groups of components that are related in a highly integrated fashion and operate with substantial independence of the rest. These may be usefully distinguished as sub-systems.

A simple, non-agricultural example may serve to convey the essential features of a sub-system, as the term is being used here.*

Suppose we think of a house as a system and wish to understand how it functions. We could try to do this room by room, considering each one in turn, and for some purposes this would be useful, but there are many connections between rooms, such as wires, beams and pipes, that are essentially related to similar wires, beams and pipes in other rooms. They are, in fact, elements of sub-systems. One way of trying to understand how the whole house functioned would be to try and understand its sub-systems. Let us take as an example the plumbing sub-system, illustrated in Fig. 2.17.

It is obvious that there are many advantages in considering such a sub-system in its entirety. Yet its components are also parts of other sub-systems within the house and in Fig. 2.17 three closely related sub-systems are shown together, concerned with water supply, drainage and central heating, respectively.

* Other workers have used the term in different ways (see Appendix).

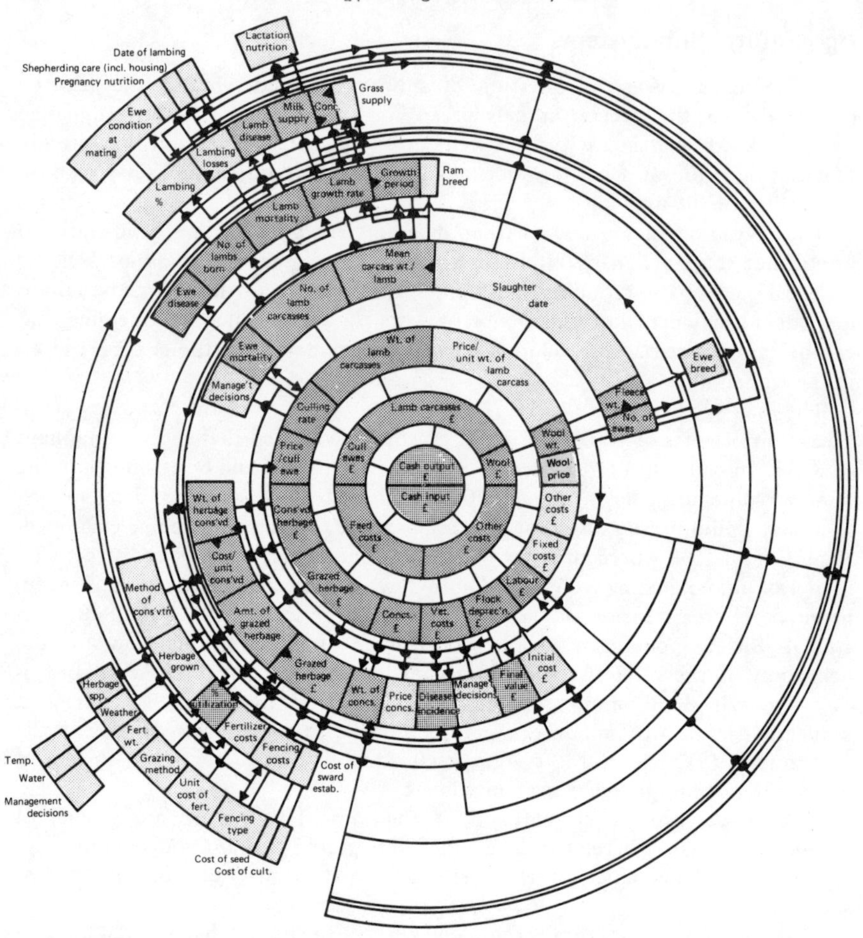

FIGURE 2.16 Components in sheep production (including all interactions).

Electrical sub-systems can also be distinguished fairly easily (a car is a good system to consider from this point of view). The point is that it is often possible to build up a better total picture of a complicated system by considering its sub-systems than by looking at parts or components. It is the relative independence of sub-systems that makes it tolerable to study them separately and makes it possible to fit the results of such studies together again, with the minimum of interactions between them.

Exactly the same argument can be applied to the study of an individual animal: that there is much sense in studying the skeletal, circulatory, respiratory, alimentary or nervous sub-systems. These examples seem more obvious, because they are more familiar, and we are really only recognizing that this is what we already do. They are also useful, however, in suggesting *how* sub-systems may be

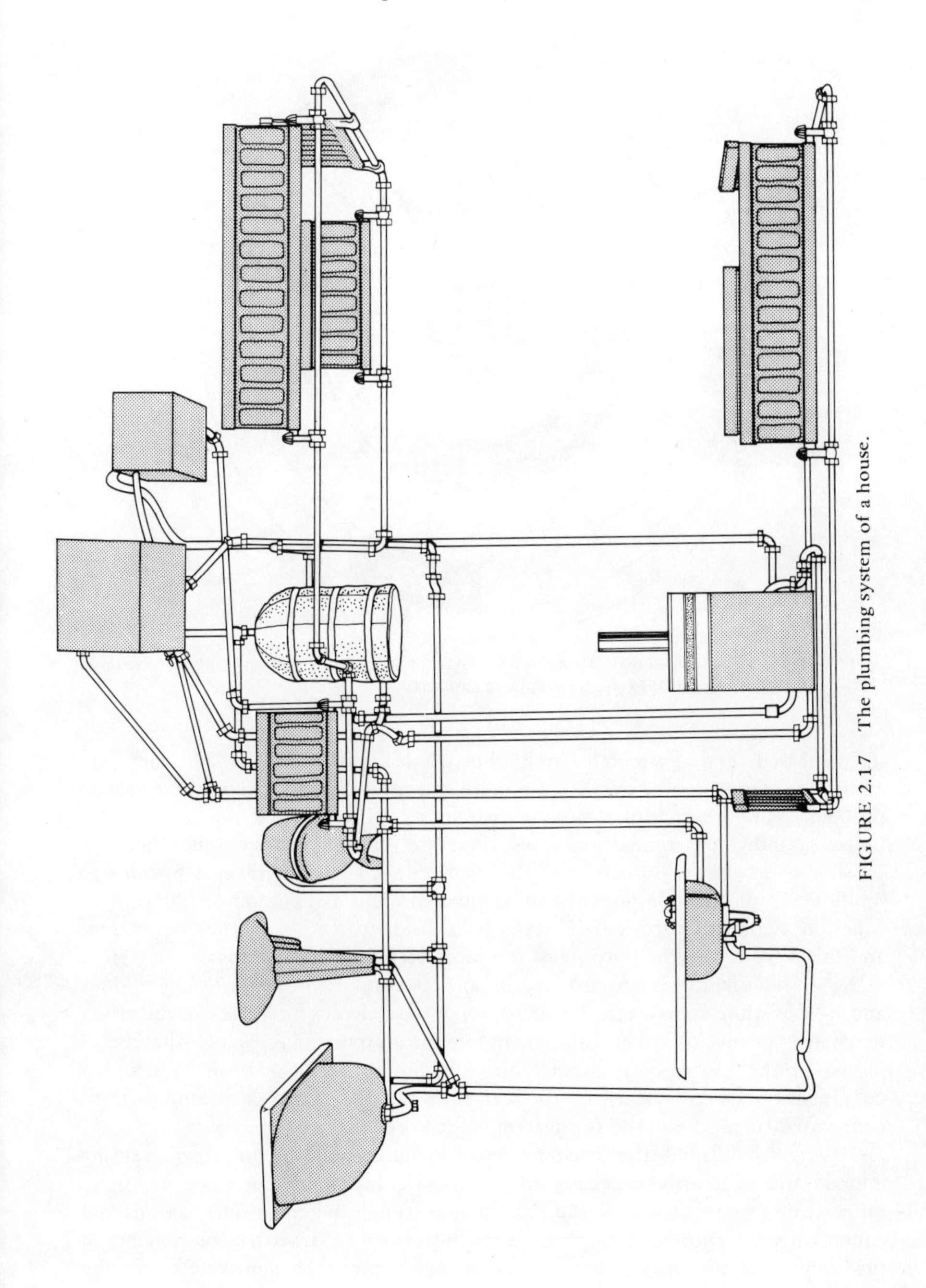

FIGURE 2.17 The plumbing system of a house.

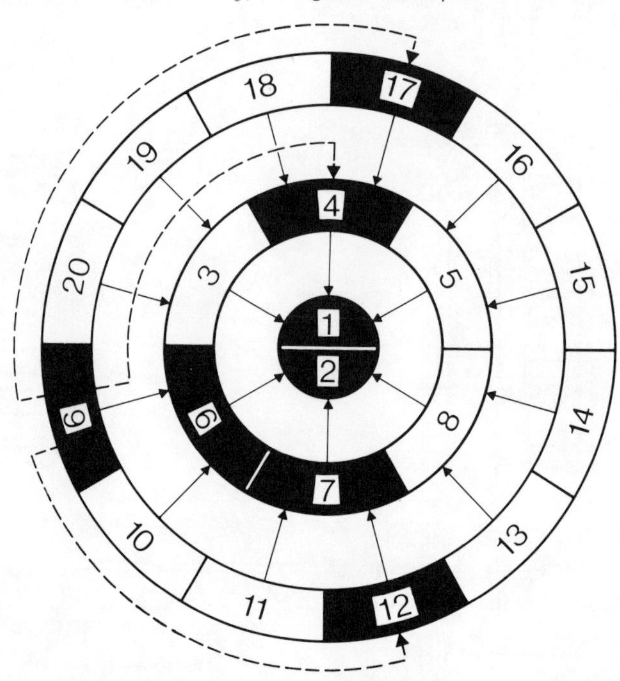

FIGURE 2.18 The identification of a sub-system in terms of its components (shown in black) and their interactions (connecting lines and arrows).

distinguished and extracted from the whole systems in which they are embedded. Once you know they are there and what they look like, there is little problem, but in agricultural systems we start without such knowledge. In the house or individual animal examples, there are physical connections: once it is possible to establish the nature of the connections that matter, it is possible to "pull out" all the components so connected. This is as much a question of function as of structure, but the essence is in identification of the connections and this is related to the purpose of the sub-system.

Now, in agricultural systems, the important connections are the interactions and it is possible to say that by definition we are always interested in the effect of change in one or other component on the output: that is, on whatever is placed in the centre of a circular diagram. Since the mechanism by which a component has its effect on the centre is by the network of interactions connecting them, this is the sub-system we require.

Having established the master-model, including all the lines representing interactions, it is only necessary to specify a component (for example, on an outer circle) and extract all the other components, linked by interactions and connecting the chosen component with the centre. Extracting components in this way does not mean that they only enter into one sub-system: on the contrary, centrally-placed components will come into most sub-systems.

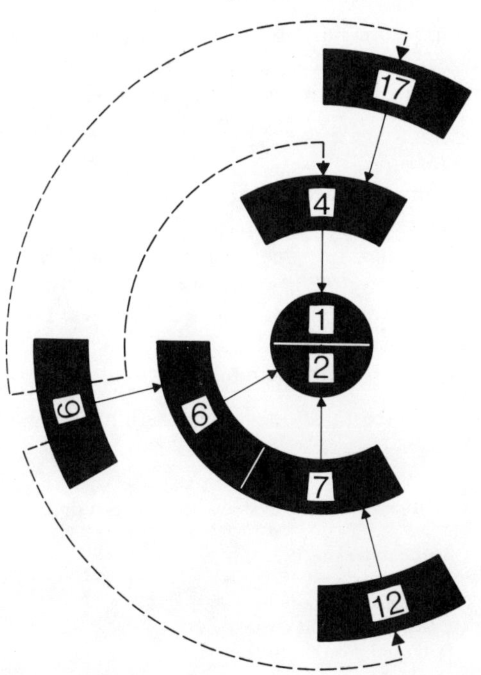

FIGURE 2.19 The sub-system shown in Fig. 2.18, extracted from the complete system. The sub-system can be defined as the effect of changes in component 9 on the ratio $\frac{1}{2}$.

Sub-systems extracted in this way (see Figs 2.18-2.19 for illustrations), however, represent the minimum that must be included in a model or a physical experiment, if all the interactions that occur in the system are to be taken into account. If it can be assumed, for specified purposes, that certain interactions can be safely ignored, then the sub-system can be further simplified, but this simplification is then recognized and the assumptions underlying it are known.

If a sub-system of a profit-centred model were studied, it would not be possible to determine the absolute profit, since other factors would influence this, but it would be possible to predict the effect on profit that changes in this sub-system would have when it was operating within the whole system.

The Description of Systems

The problem with all descriptions is knowing where to stop. The amount of detail required varies with the purpose of the description: if the purpose is to identify an individual it has to include a feature or combination of features that are unique to that individual. One purpose in an agricultural systems context is to make it possible to recognize a member of a classification group. It is necessary, therefore, to include essential, diagnostic features and to exclude trivial detail. Another important purpose is to understand how a system functions and a third is to make it possible for someone to copy it.

The most useful description is one that will serve all these purposes and this means describing:

(a) diagnostic, essential features;
(b) features or values that are essential but which may vary within stated limits without making it a different system;
(c) features that are essential but may vary very widely indeed: the permissible variation (in, say, oxygen content of the air) may be greater than is ever likely to be encountered and such features, therefore, need never be mentioned;
(d) trivial or unimportant features that should be omitted.

Those features that are included in (c) may most usefully be regarded as part of the *context* in which a system operates.

TABLE 2.1 Characteristics of Grazing Systems[a] (from Spedding, 1973a)

Character	Value to be specified
Plant population	Species and variety (age etc.) Density (per unit area) Potential growth curve or yield.[b]
Animal population	Species and breed Size, age, sex etc. Number (per unit area) of each type. Nutritional requirement (total and seasonal distribution).
Grazing period	Continuous, on/off, seasonal (with dates).
Non-grazing period	Details of treatment of both plants and animals when grazing is not taking place.
Grazing method	Rotational, strip, continuous etc. (and periods of use). Dimensions of paddocks, strips etc. Frequency of rotation etc.
Conservation	Proportion of area (or area/days) conserved. Method(s) of conservation. Dates and quantities.
Imported foods	Timing, level and nature of supplementary feeding.
Disease control	Control methods applied (drugs, vaccines etc.). Timing and quantity.
Layout of grazing area	Including watering and feeding arrangements.
Labour requirement	Profile of seasonal distribution: quantity and skills needed.
Key husbandary features	Especially at critical times (e.g. parturition).

[a] visualized as operating in a variety of environments, which also have to be closely specified (e.g. in terms of nutrient status of the soil, rainfall etc.).

Thus, grazing pressure is regarded as the result of interactions between a grazing system and the environment in which it is operated. The same applies to actual yield of herbage: it is only the *potential* yield[b] that is characteristic of the grazing system.

System description is thus chiefly concerned with (a) and (b) but these "features" may be components, relationships, factors or any of the items shown in Fig. 2.11.

It is obvious that system description requires a great deal of knowledge about the system and the way it works. It is a vital procedure, however, and methods are as yet poorly worked out. It would be a great advantage to be able to fit descriptions into some systematic pattern, so that the procedure would be like filling up a form. This would make it much easier to compare descriptions, for example. But a word-description is unlikely to be sufficient by itself and it is better to think of a form combined with some pictures.

The first set of pictures to be considered are the circular diagrams and their derived sub-systems. These summarize the entire system over a period of time but they are not dynamic: they do not really take time into account and therefore convey a very static picture. Yet the operation of an agricultural system is essentially a linked set of processes and timing is often of critical importance. Table 2.1 shows the essential information required for an agricultural grazing system and Fig. 2.20 illustrates how periods of time can be related to some of the items in a sheep production system.

The progress of different operations can be charted in various ways and the timing relations between them can be shown as a "critical path" diagram. Cash flow and capital requirements can usefully be depicted (see Fig. 2.21), as can the flow of physical resources and the distribution of demand for labour (Fig. 2.22). Clearly, the number of possible, and useful, pictures is very large and it would be nice to have all the information in one diagram.

It is important to recognize that this is not possible, simply because one picture is inadequate to describe even one simple object: a complex system will never be conveyed by one picture alone. In case this sounds rather sweeping,

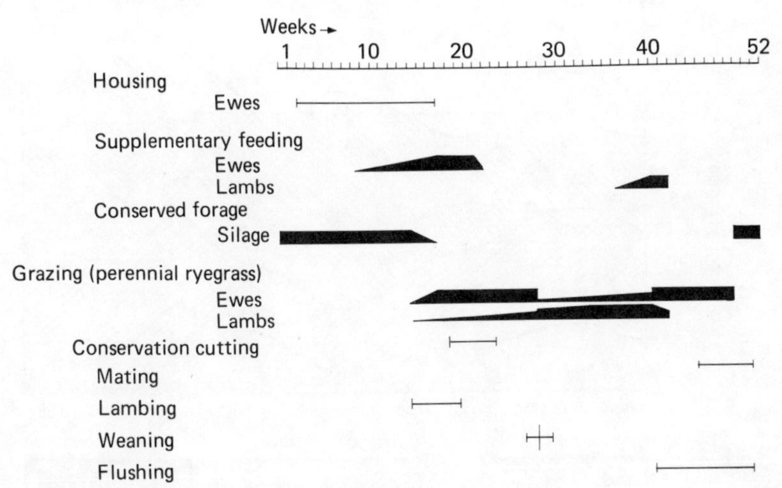

FIGURE 2.20 A diagrammatic representation of facilities required and timetable of main events in a sheep production system. The thickness of the lines can be used to indicate quantities (from Spedding, 1972a).

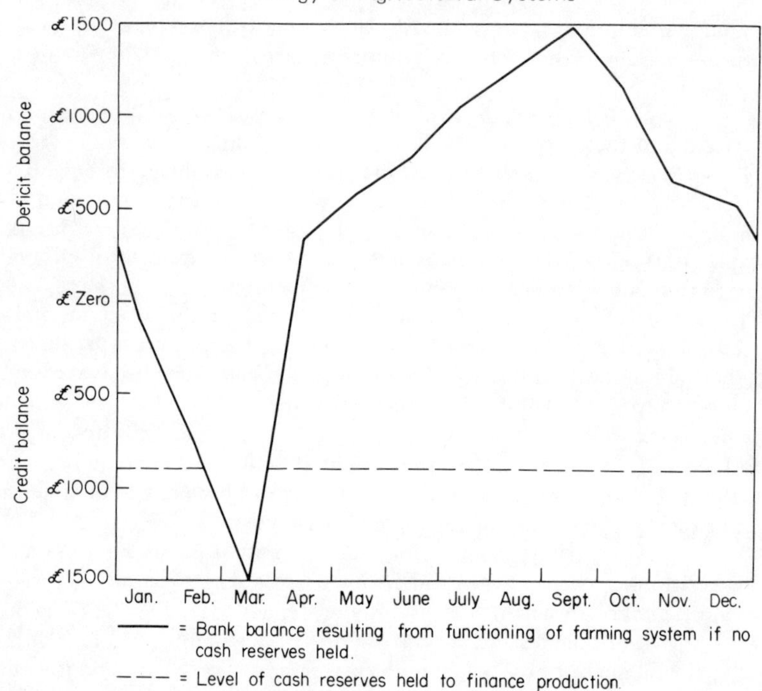

FIGURE 2.21 Seasonal capital requirements of dairy and mixed arable farm (from Lloyd, 1967).

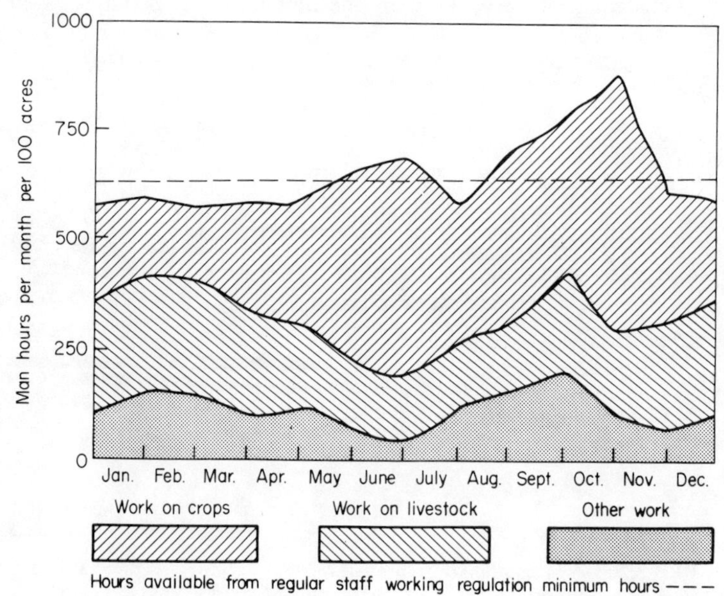

FIGURE 2.22 Seasonal labour distribution of arable-with-cash-roots, beef-and-sheep farm (fom Lloyd, 1967).

consider a horse. Because we already know what a horse looks like, it seems easy to provide a picture that will describe it. Most pictures, however, will show only the outside of one side of a stationary horse. We assume that it has other sides, that it moves in some fashion (and that it eats, breathes and so on), and that it has an inside, but the picture describes none of these attributes.

It is interesting to consider how it is that we manage to achieve a more complete, better-rounded picture of the whole horse in our minds. We do it mostly by a mental "running round" the horse, rapidly reviewing a whole series of pictures and, if we try to combine too many, the result is confusion. A ciné film achieves a comparable effect in relation to dynamic situations.

Similarly, whilst the circular diagrams may serve as a "master-plan" relating all aspects, a range of different supporting diagrams is needed to describe a system in the detail that is wanted. Some "minimum album" needs to be worked out for each major system category, in order to systematize description, bearing in mind that description, like classification, varies with purpose. One special and very important purpose relates to the choice of system.

The Choice of Agricultural System (see also Chapter 8)

In many cases, the choice has already been made and only minor changes are possible. In essence, however, this is not very different from a situation where we start with no agriculture at all. The fact is that the choice has to be made in the light of the resources that exist and the costs of obtaining inputs. If there are substantial natural or existing resources, it makes some sense to try and match

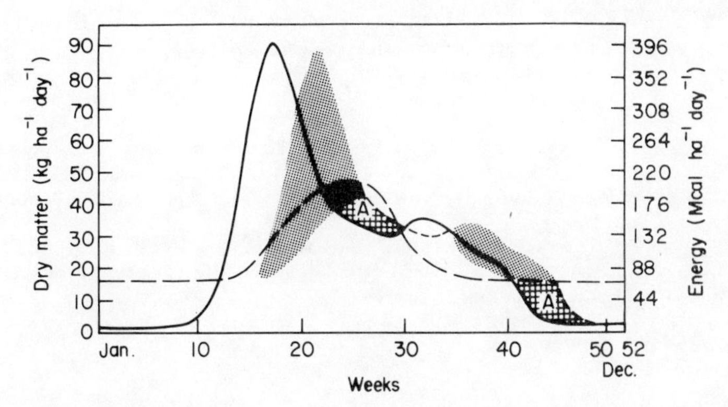

Herbage surplus to current requirements, that could be available 4 weeks later.

Deficiencies in sheep requirements that could be satisfied with herbage remaining *in situ* for 4 weeks.

The rest of the excess would be available for conservation (from Spedding, 1972b).

FIGURE 2.23 The availability of herbage for harvesting. Herbage production curve (————) to give 38,812 Mcal/year. Sheep requirement curve (– – – – –) for 12 ewes with singles, requiring 38,472 Mcal/year.

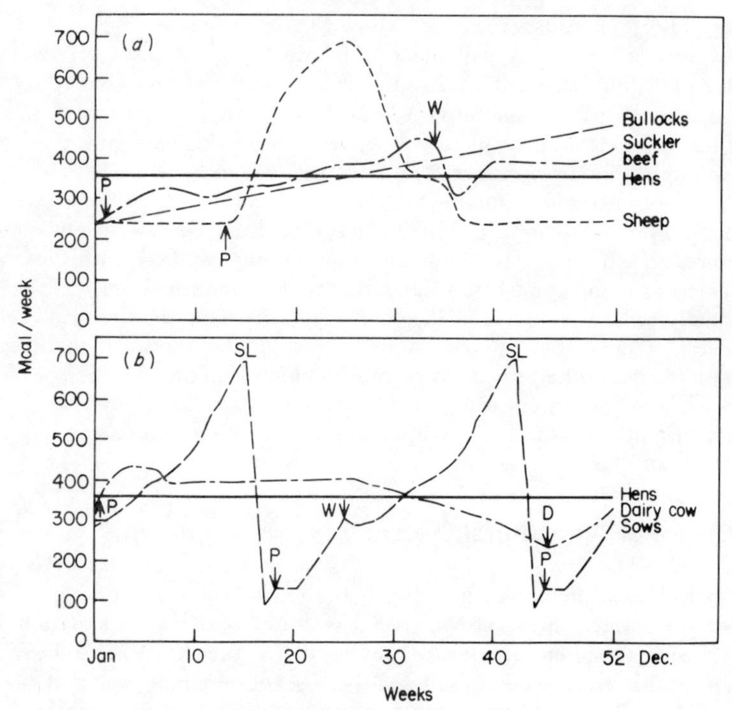

FIGURE 2.24 *Patterns of energy requirement for different species* all adjusted to the same annual requirement of 18 500 Mcals.

(a) (b) ——— 127 laying hens (theoretical calculation from data supplied by T. R. Morris, University of Reading).

(a) – – – – – 1.8 Bullocks from weaning to slaughter at 14 mths (from data supplied by J. C. Tayler, G.R.I.).

(a) –·–·–· 1.2 Beef suckler cows and calves. (Data supplied by R. D. Baker, G.R.I. H.768, based on 11 cows [mean wt. 450 kg] with single calves feed indoors. Four of the cows were in calf again approximately 3 months after calving.)

(a) ------ 5.8 Ewes with single lambs. (Data from R. V. Large, G.R.I. H.683.)

(b) –·–·–· 1 Dairy cow. Theoretical calculation for a Friesian-type cow giving 1200 gal of milk/annum.

(b) – – – – – 1.1 Sows having 2 litters of 8. Theoretical calculation.

P–parturition
W–weaning
Sl–slaughter
D–dry (i.e. non-lactating) [from Spedding, 1972b].

the needs of the agricultural system chosen to the available resources, as closely as possible. The problems and possibilities of matching patterns of resource requirement and availability are illustrated in Figs 2.23 and 2.24. The problems are greatest where the natural resources show marked variation (or, even worse, unpredictability) in items (such as temperature) that are difficult or costly to support or manipulate by increasing inputs.

The appropriate descriptions for these purposes rely heavily on conveying patterns but risks and uncertainties are also of great consequence. Averaging patterns is apt to disguise variation, but description has also to cope with this difficult problem, for variation is a characteristic of the biological systems on which agriculture is based.

References

Dale, M. B. (1970). *Ecology* **51**, 2.

Edmond, J. B., Musser, A. M. and Andrews, F. S. (1957). "Fundamentals of Horticulture" (2nd edition), McGraw-Hill, New York.

Garner, F. H. (1972). *In* "Modern British Farming Systems" (F. H. Garner, ed.), pp. 203-204. Paul Elek, London.

Lloyd, D. H. (1967). *O & M. J.* **4** (3), 1-8.

Meadows, D. H., Meadows, D. L., Randers, J. and Behrens, W. W. (1972). "The Limits to Growth". Earth Island, London.

Ruthenberg, H. (1971). "Farming Systems in the Tropics". Oxford University Press.

Schery, R. W. (1972). "Plants for Man" (2nd edition). Prentice-Hall, New Jersey.

Spedding, C. R. W. (1971). *Outlook on Agriculture* **6** (6), 242-7.

Spedding, C. R. W. (1972a). *Fourrages*, **No. 51**, 3-18.

Spedding, C. R. W. (1972b). Proc. 2nd Wld Congr. Anim. Feed., Madrid, 1972, 4, 1245-58.

Spedding, C. R. W. (1973a). Proc. IIIrd World Conf. Anim. Prod. Melbourne. 2.1-2.20.

Spedding, C. R. W. (1973b). Proc. Symp. "Modern Methods of Sheep Production", Green week exhibition, Berlin, 1973.

Spedding, C. R. W. and Brockington, N. R. (1974). *Agric. Syst.* (In Press).

3 Biological Systems

The main difference between agricultural and biological systems is that the relevant criteria for assessing them are essentially different.

Agricultural systems may be judged in relation to any of the purposes for which they are carried out, but any complete appraisal must include economic assessment. To make this possible, it is necessary to include in an agricultural system everything that contributes to costs or to financial output.

Biological systems, on the other hand, whilst there is nothing wrong with costing them, must essentially be judged by biological criteria. This may often include the output of agricultural products, but the latter are still components of the biological system considered.

Sometimes an agricultural system consists of a biological production process with additional non-biological components of economic significance. In other cases, there are additional biological systems involved that interact with but are not the same as the main production process. They may be of great importance but they are not central to the study of agricultural systems: some of them are peripheral to any agricultural interest but of concern to the community (see Chapters 8 and 9).

Of course, many biological processes can be described as systems but those most relevant to agriculture fall into two main categories that may conveniently be called production systems and component systems.

2. Production Systems

These are concerned with the output of agricultural products of biological origin, food for man and livestock being notable examples. There are thus as many production systems as there are products and ways of producing them. As with agricultural systems, such a large range of diverse systems can only be dealt with by some form of classification. Fortunately, since this category of biological systems consists of those biological processes upon which agricultural systems are based, the same kind of classification will serve, except perhaps for the finer sub-divisions.

It is better, therefore, to use the same classification wherever possible. A production system can then be characterized chiefly in terms of the underlying biological processes involved in the production of a specified product (e.g. beef)

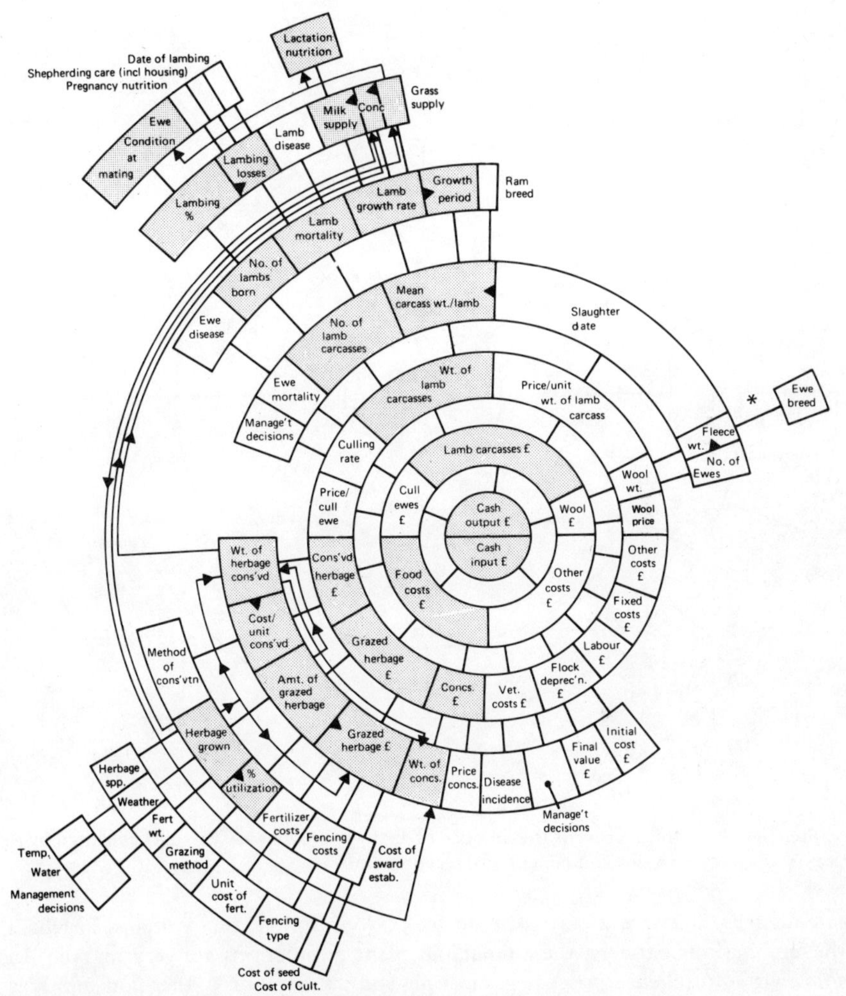

FIGURE 3.1 Effect of herbage grown on financial outcome of sheep production (*all such radial lines represent inwardly-directed arrows).

from a specified population (e.g. suckler cows of the Hereford x Friesian breed mated to a South Devon bull), breeding at a specified time (e.g. autumn) and fed on a specified diet (e.g. mainly grass).

The same kind of circular diagrams can be used to work out the main components and their interactions and the same sort of problems occur in trying to describe the systems identified.

The same approach may also be used to try and extract relatively independent sub-systems: an example of this is shown in Fig. 3.1. This is

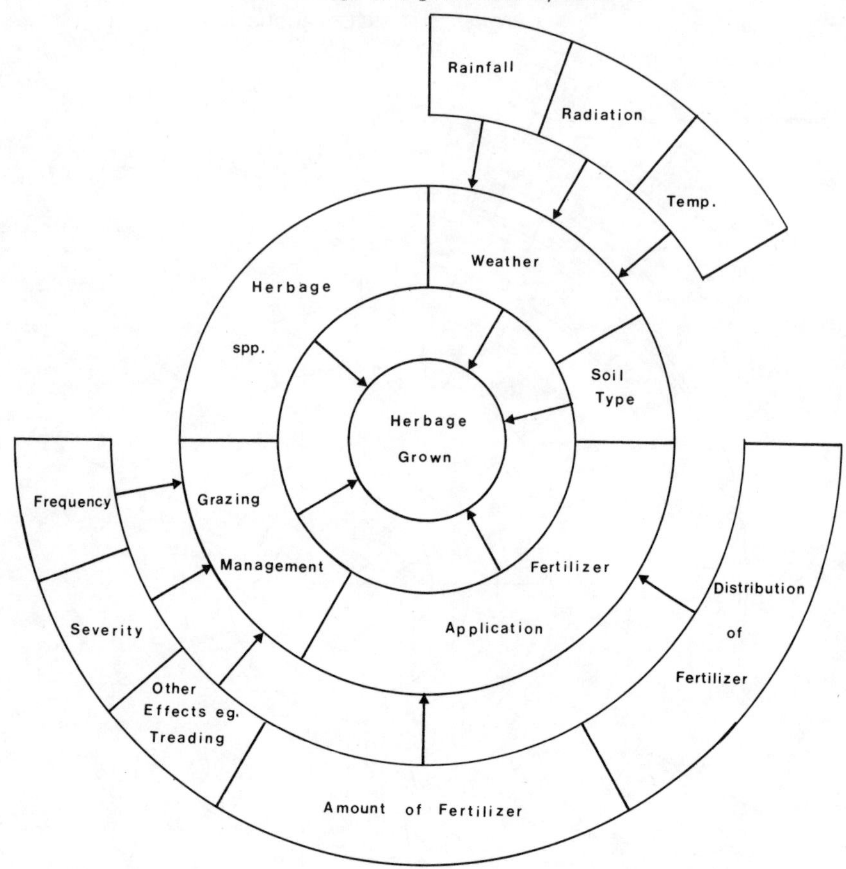

FIGURE 3.2 Influences on the amount of herbage grown (see Fig. 3.1) as shown by a circular diagram with this component at the centre.

intended to illustrate a particular difficulty found in grazing systems, in which the interactions between the animal and plant populations are very marked. In some agricultural enterprises (e.g. most pig and poultry units), the food supply is under the control of the operator and is unaffected by the behaviour of the animals being fed. In the grazing situation it is quite different. The numbers, size, weight, movement and grazing activity of the animals may not only greatly affect the amount of herbage harvested but may also influence markedly the amount of herbage grown. Furthermore, these effects are modified by the nature and condition of the soil and thus by the weather.

Superficially, one of the most readily extracted sub-systems is that concerned with growing the herbage, regarding the latter as an intermediate crop and the animal utilization of it as another sub-system. Unfortunately, these interactions between animal and plant populations may render this invalid, simply because some of the factors affecting plant growth are omitted with the animal. Figure

3.1 deals with this problem for the case where we wish to vary a peripheral factor and study its effect on an intermediate product. The same thing can be done, perhaps more clearly, by putting the intermediate product at the centre of a separate diagram (see Fig. 3.2).

This illustrates one of the difficulties of building up a picture of a system from studies of components. In a sense, this is what classical research tries to do. It assumes, rightly, that the whole system is too complicated to understand by studies of the whole system alone: it further assumes that the remedy is to study the components separately and then put them together again. In the latter notion, as discussed in Chapter 2, there often lies a considerable oversimplification. It is true that much appears to depend upon definitions, especially of "component", but the essential point remains, that knowledge of components is insufficient without studies of all their interactions when combined.

So the objects of studying production systems and component systems may be quite different. In the latter case, the primary objective is to improve our knowledge about the component system and how it works (e.g. what governs herbage growth and production). In the former case, however, the main object is

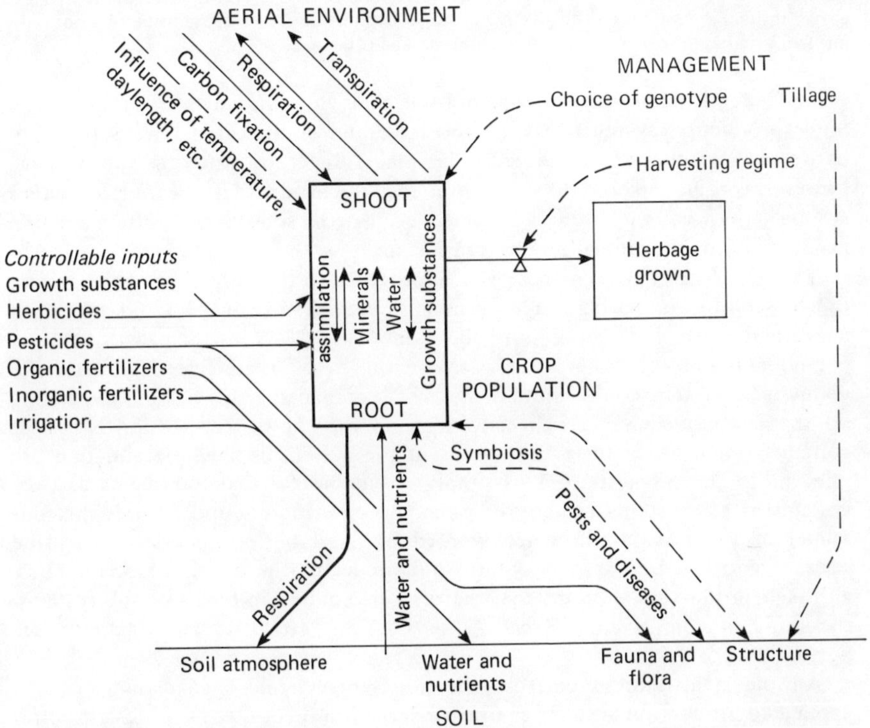

FIGURE 3.3 The main flows and components in the growth of herbage, to illustrate the kind of factors apparently involved when herbage growth is viewed in isolation.

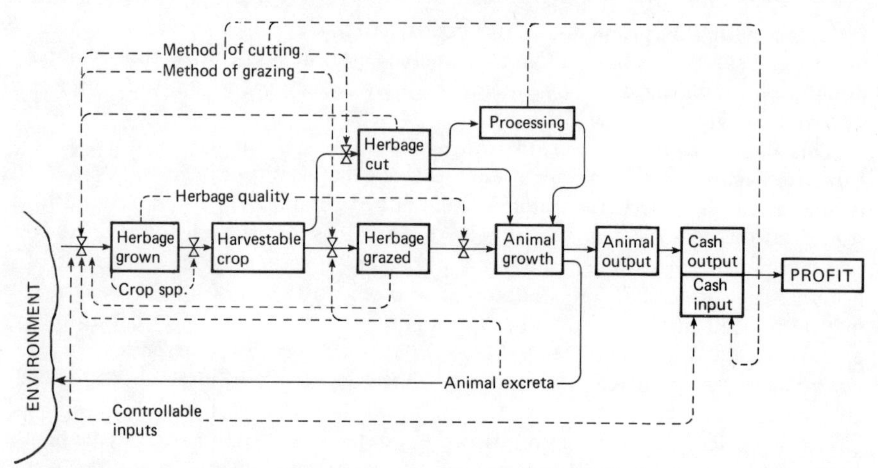

FIGURE 3.4 Herbage growth shown as a component of the larger system. As a sub-system it has connections all the way to the system output, some of which have feedback effects on herbage growth that were not recognized in Fig. 3.3. For example, it is necessary to specify the grazing animals involved before it is possible to include their influence on the amount of herbage grown. Thus, many of the events after the herbage is grown influence its growth and all the controllable inputs and management factors not only influence herbage growth but also contribute to the total costs.

to understand the production system itself, and what influences its productivity. Since production systems may be too large and complicated always to study entire, sub-systems are extracted. The main object of studying sub-systems, however, is to understand how they influence the productivity of the production system (that is, e.g. how the herbage growth sub-system influences the productivity of the animal production system.)

The difference between these two approaches to the same part of the system can be seen by comparing two traditional flow-diagrams (Figs 3.3 and 3.4), both concerned with herbage growth but one (Fig. 3.3) dealing with it as a component and the other (Fig. 3.4) treating it as a sub-system. Clearly the component system could contain more or less than the sub-system, depending on the context envisaged. The advantage of the sub-system version is that it contains only the features that occur in the system as a whole and that are relevant to the sub-system. This simply re-emphasizes that looking at parts of systems as sub-systems retains the same general structure and ensures that our model of the sub-system, when worked out and better understood, will be relevant and can be fitted back into our model of the original system. Thus, although production systems may have to be studied in terms of sub-systems, the study of components is really a separate and rather different activity (see Section 2 of this chapter).

Although the same broad classification scheme, and even naming, can be applied to production systems as to their agricultural counterparts, description is a different matter. This is because the main interest and purpose in considering them is also different.

The problem with agricultural systems is that they must be expressed in relevant economic terms but they cannot be wholly understood in these terms. Part of the understanding required in order to manipulate agricultural systems successfully is that of the underlying biological systems (the same argument applies, of course, to other underlying sub-systems, involving money or labour, for example).

So the main object of considering a production system is to understand how it works and what factors influence the biological processes involved. Description is therefore concerned to portray processes rather than static structures and this is one reason why flow-diagrams have been frequently used.

Flow-Diagrams

It is not entirely true that all biological processes consist of, or can be adequately represented by, flows, although it may be so if information flows are included as well as those of materials. However, a biological system can no more be completely represented by any one picture than can an agricultural system. Part of the difficulty with a production system is that the product, by definition the most interesting feature, does not flow but is formed, from flows of many substances, and subsequently "extracted".

It is worth considering whether *all* the relevant flows can be represented, since any one is bound to be inadequate except for special purposes. Clearly, if the main interest in a system is in the energy transformation or the formation of protein, then energy and nitrogen flows would be relevant. The same argument applies to water, dry matter, cadmium or salt, provided that one of these constituents is specified as the main interest.

This is quite different from arguing that the flow of any one of them will adequately represent the functioning of the whole system, however. It has been argued that if only one expression has to be chosen it should be energy, because it can be regarded as the driving force of all living systems and because it cannot be recirculated (Lindemann, 1942; Macfadyen, 1957; Slobodkin, 1959; Phillipson, 1966): each unit of energy is therefore used once only, although energy can be stored and such stores transported within the system.

However, the value and significance of the product may not be related to its energy content, and productivity may not reflect the "activity" of the system as represented by the rate of energy flow through or within it. Increasing the stocking rate of grazing animals, for example, can be taken to the point where the quantity of herbage energy being ingested is maximized but where, as a result, each animal only receives enough for maintenance (or less) and there is no animal production at all. Nor is energy the only determinant of production at any stage and an energy flow-diagram often shows little of the dependence of this flow on the flows of other materials (e.g. N and water). Many of the other resources required are costly and scarce and may be used very wastefully if provided in excess: it may frequently be necessary, therefore, to increase the efficiency with which they are used, even at the expense of energetic efficiency.

Even in purely biological terms energy content is not a completely satisfactory expression, since the functions and significance of animals and plants (and parts of them) are not necessarily related to their energy content at all. This is not just a problem of systems description, however; it is applicable to many aspects of biology. Exactly the same kind of difficulties occur in describing growth in plants and animals (see Chapters 4 and 5).

It is easy to point out the deficiencies of any description based on one constituent but it is important to recognize that a description in terms of *all* constituents is not only impracticable but also undesirable. Descriptions, like other models, have to be essential abstracts of the real thing. If they are no simpler than the real thing, they generally serve little purpose and are no easier to understand. What is needed is a simplicity that reduces systems to their essentials. So we should be asking, for some specified purpose, what does this essential simplicity consist of and how many constituents must be included?

Now it cannot be entirely known, in advance, what the essentials are, or it would not be necessary to engage in further study. The object is to express our understanding in as simple a fashion as represents it adequately, and then to compare this model with reality; in this way, our degree of understanding is tested. One way to explore this line of argument is to see whether it is possible to describe a system in terms of *several* obviously important constituents, simultaneously. This is attempted in Fig. 3.5, for energy, nitrogen and water in the production of wool per unit of land. Every line with an arrow represents a flow of one of these three substances: each flow occurs at a rate that is dependent upon a great many factors. None of these details is shown in the

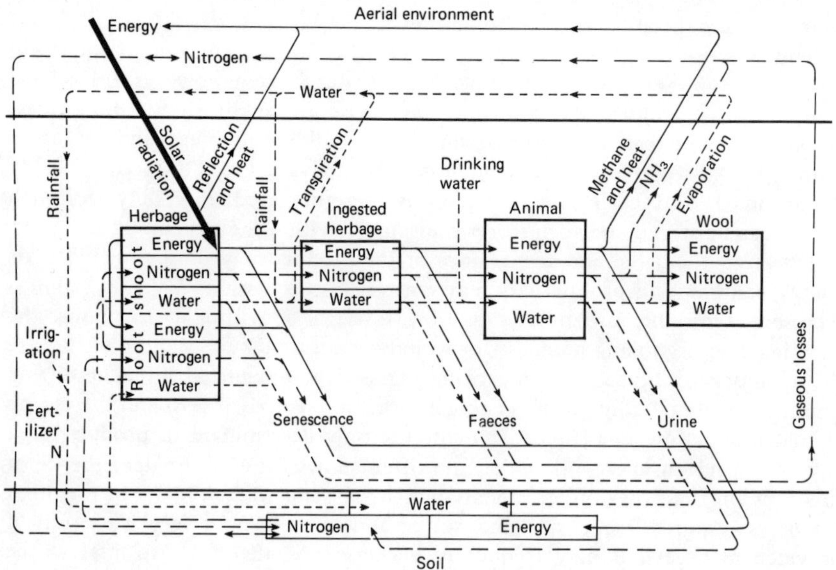

FIGURE 3.5 The flow of energy, nitrogen and water in wool production (Spedding, 1973).

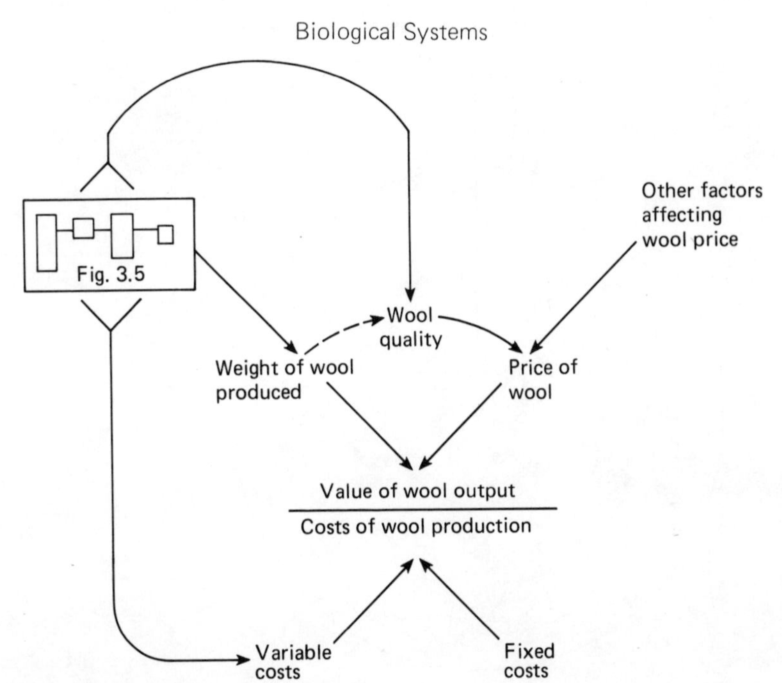

FIGURE 3.6 The relationship between Fig. 3.5 and the agricultural system of wool production (Spedding, 1973).

diagram, which is nonetheless complicated enough. The quantities (or levels) of each substance in herbage, animal, soil or aerial environment, may influence the rates of flow of the same or any other substance. The complexity for even three constituents in a system chosen for its simplicity, imagining one kind of animal and one kind of plant, is considerable.

Skill in constructing and in the actual drawing of diagrams can help greatly: the use of shapes, colours and the choice of angles are all vital in allowing a diagram to clarify rather than confuse. It is worth *stating* the complexity diagrammatically, but it is infinitely better to convey a clear picture of it. A further step would be to add a third dimension to keep different information separate yet connected. Figure 3.6 is a simple representation of the way in which Fig. 3.5 fits into the agricultural system of wool production.

Mathematical Models

One very useful function of flow-diagrams is as precursors of mathematical models. This is a large, rapidly developing subject with much to offer the biologist, but it can only receive brief mention here (for further details see Watt, 1966; Jones, 1970; Lieth, 1971; Brockington, 1971a; Patten, 1971).

Briefly, the object is to quantify the relationships shown in a flow-diagram, in a form suitable for incorporation in a computer program. It is then possible to

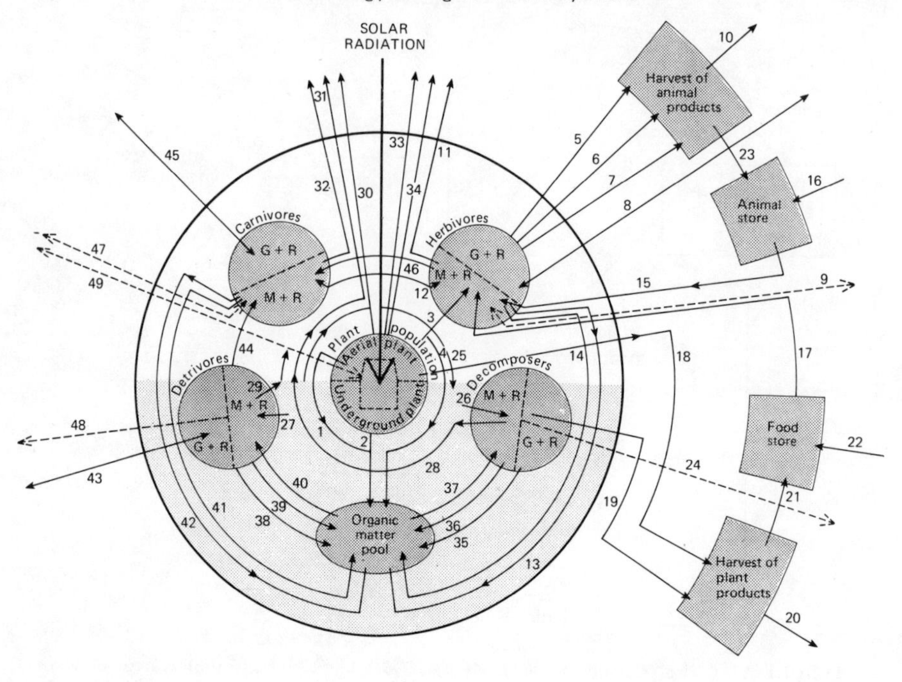

Key to Line Numbers

Line Number	
1	Dead material
2	Dead roots
3	Grazing
4	Dead material (fouling, treading, etc.)
5	Milk
6	Wool
7	Whole animal
8	Emigrants/immigrants
9	Gaseous losses and inputs
10	Consumption of animal products
11	Heat losses
12	Heat
13	Excreta
14	Dead matter
15	Replacement stock
16	Purchased stock
17	Conserved feed
18	Cut herbage
19	Mushrooms (e.g.)
20	Sale of conserved feeds
21	Storage of conserved feeds
22	Purchase of conserved feeds
23	Retention of stock
24	Gaseous losses
25	Heat
26	Heat
27	Heat
28	Heat
29	Heat
30	Reflection
31	Conduction
32	Heat losses
33	Heat (respiration)
34	Heat (transpiration)
35	Excreta
36	Dead matter
37	Food source
38	Excreta
39	Dead matter
40	Food source
41	Dead matter
42	Excreta
43	Emigrants/immigrants
44	Consumption of detrivores
45	Emigrants/immigrants
46	Consumption of herbivores
47	Gaseous losses of input
48	Gaseous losses
49	Gaseous input

calculate the consequences to the whole system of manipulating the values attached to any of the component levels or rates. This procedure allows theoretical experiments to be carried out over a range of values that could never be covered by physical experimentation. Since the emergent conclusions and hypotheses need testing, physical experimentation is not rendered unnecessary but it is possible to choose the more interesting or promising lines of enquiry and thus to make better use of limited experimental facilities.

There are, of course, many different ways of constructing what are essentially flow-diagrams: one of them is illustrated in Fig. 3.7, representing an energy budget.

Energy Budgets

Budgets can be calculated, without diagrams, but it is often helpful to show visually "where the energy goes" (or, indeed, where any other constituent goes) in a production system. Whether calculations or diagrams are used, there are the usual problems of what to put in and what to leave out, how finely to sub-divide categories and components, where to place the boundary of the system and how to define the context. Figure 3.7 shows a definite boundary and some items outside it.

Now it may be argued that if items are important enough they should be included in the system: especially if any kind of "feedback" mechanism is involved. In fact, the "livestock" and "feed-pools" (or reservoirs) shown are good examples of what might be called controlled feedback mechanisms. There is little doubt (as argued by de Wit, 1972) that if an output of a system has some automatic consequence that affects the system, it is better to include all of this within the system boundary. Thus if a pig is thought of as a biological system, the appropriate boundary is near the skin, perhaps including a layer of air and, with it, a micro-climatic zone (see Fig. 3.8). The heat output of the pig is small in relation to the total atmosphere and has a negligible effect on it; there is therefore no feedback of raised environmental temperature on the pig and no effects on its heat production or anything else. However, if we imagine the same

---

FIGURE 3.7 A diagram illustrating the way in which energy supplied by solar radiation, to a U.K. agricultural grazing system, is dispersed through various routes (from J. M. Walsingham, 1974). The quantities of energy (in Mcal) stored, at a point in time during spring, in the main components are given below:

Plant population	78 000
Herbivores	
(a) domestic	7634
(b) non-domestic	100
Carnivores	2.7
Detrivores	1055
Decomposers	3.5

M + R = Maintenance + Rejection
G + R = Growth and Reproduction.

The numbers associated with the arrows are referred to in the key.

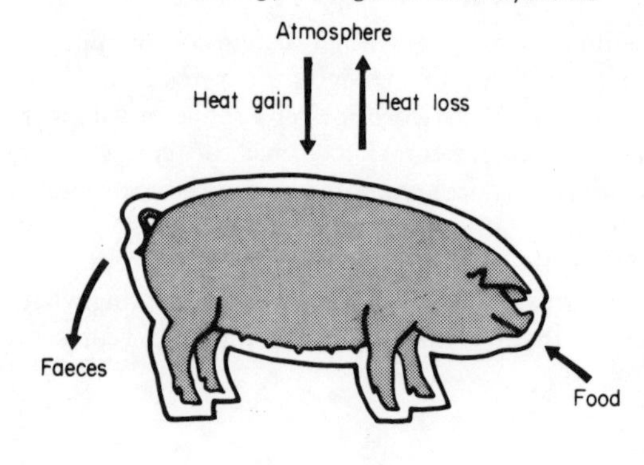

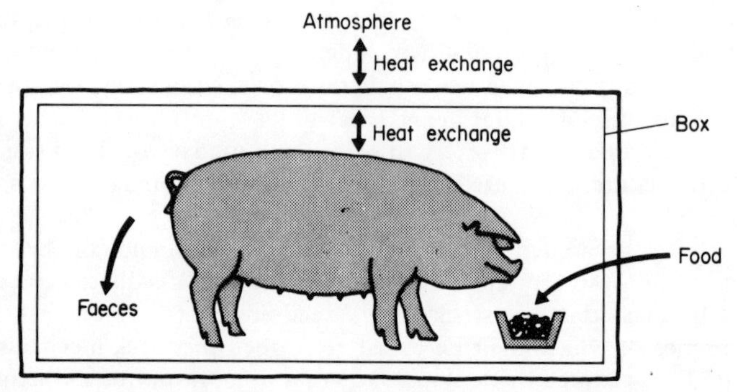

FIGURE 3.8 The effect of physical features (in this case a box) on the position of the system boundary (shown as a heavy line); the repositioning is required because of the feedback between heat from the pig to the box atmosphere and back again.

pig in a relatively small house (a kind of box), the heat output of the pig will raise the environmental temperature and there is an automatic (uncontrolled) feedback to the pig. In these circumstances, it is better to regard the outside of the box as the boundary of a "pig-in-box" system.

In Fig. 3.7, the outputs of livestock and feed could be sold and need never enter the system again. On the other hand, the system may require the replacement of breeding stock or the importation of additional feed and these could come from the reserves to which the system has contributed, or they might be the very same animals and feed originally supplied to the pool, or they could be obtained from entirely independent sources. Someone outside the system, in this case a farmer, decides which way suits him best and operates as a controlling mechanism. If the same animals or feed do return to the system, however, they may operate in a "feedback" fashion and influence the system in a special way because they were produced within it: they might have some peculiar properties or deficiencies, for example. But since it is not automatic, it is best regarded as a mechanism of controlled feedback and, because it may operate in this way, it is worth including in a picture of the system.

Pictures based on flows, of course, tend to obscure some of the features of biological systems that are characterized by "number"; the number of discrete individuals being, for example, one measure of the population.

Population Changes

Animal and plant populations can, of course, be represented by quantities of energy or nitrogen but the same quantity divided up into a lot of small

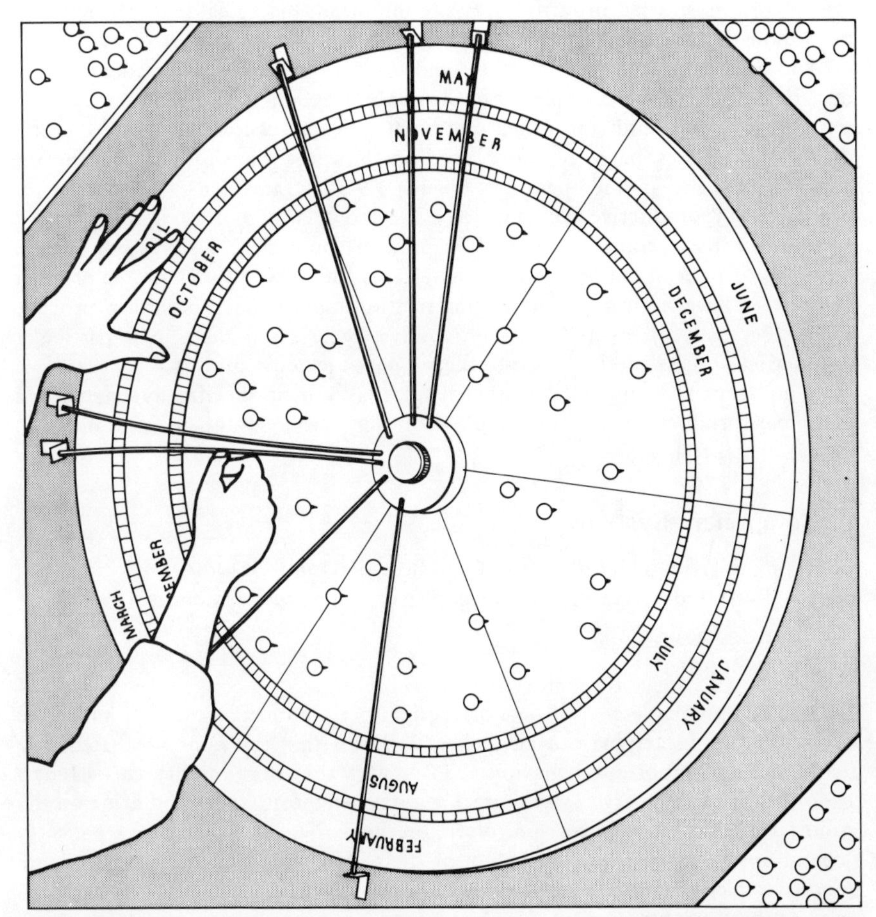

FIGURE 3.9 A circular chart used to record events in the reproductive cycles of a herd of sows. The circles represent 6-monthly periods and are segmented into months and days. Fixed radial wires mark off important landmarks, such as first oestrus, mating, weaning and veterinary treatment. Coloured pins are numbered to identify individual animals in relation to the above landmarks (with wires of the same colour as the pins).

In operation, the circular calendar is rotated each day, beneath the fixed wires, and pins are changed as events occur. [This device was developed by the *Farmer and Stockbreeder* from an original model made by the NAAS Dairy Husbandry Specialists at Bristol].

individuals or a few large ones may totally alter the biological picture. This is especially true in relation to reproduction, which is after all a major attribute of living things, since not all animals or plants are in a reproductive state. Perhaps the most striking example is to consider how different is the biology of the situation if a given amount of energy represents two individual animals (a) of the same sex, or (b) of opposite sexes.

Population changes, in numbers and structure, the nature of individuals and their physiological states, and the interactions between them (e.g. a lactating female and its suckled progeny) are very important biological features that may need to be described and pictured in quite different ways.

Another feature of reproduction, and often of growth, is a seasonality that may be unrelated, for example, to food supply. Day-length is often implicated in reproduction and temperature may dominate the expression of growth. Pictures of these processes may therefore need to show the time of year and/or the patterns of light and temperature. Figure 3.9 illustrates a useful diagram for dealing with reproductive events which cycle and occupy more than one year.

Many of these considerations also apply to component systems but for the purposes of understanding production systems they have to include the product or at least have a clear, unambiguous relationship to the production process. Thus, Fig. 3.9 relates to the reproductive cycle of animals in a particular production process and is related to the timing of product output.

Component systems, on the other hand, may not be specifically related to a particular production system, even when they are relevant to a production process in general.

2. Component Systems

The distinction being made between these two kinds of biological system can best be illustrated by taking some examples of component systems.

a. Dung-Beetles

Grazing animals are of major agricultural importance and have the characteristic of carrying out all their activities within the same area. It is true that there are important behavioural influences that may modify this (Hilder, 1966) but, to a very large extent, grazing animals deposit faeces and urine on the ground on which they graze and, often, on the herbage they eat. Many parasites depend upon this for the completion of their life-cycles, and the latter provide good examples of component systems (see d.).

If faeces were to remain on the ground and simply accumulate, herbage growth would cease as the foliage became buried and, before this, grazing would be greatly reduced because animals dislike such soiled herbage. The dispersal and decomposition of such material are therefore vital and the *rate* at which these processes occur can be very important. A component system of "fouling" of herbage and its effect on the grazing animal can be envisaged and, indeed, this has been the subject of flow-diagrams and mathematical models (Brockington, 1971b).

One component of this system consists of the fauna that influence dispersal and decomposition: they include the grazing animals themselves, dispersing it with their feet, birds searching for insects (this may result in an astonishingly wide-flung dispersal of, for example, cattle faeces (Clements, 1973), which tend to be deposited in discrete "pats"), earthworms dragging material underground, flies that breed within the faeces, and dung-beetles. The latter have been shown to be very effective in South African range-lands (Gillard, 1967) and their relative absence from Australia has been shown to be associated with the surface accumulation of dried faeces: the latter is deemed sufficiently serious to make it worthwhile importing the missing species and distributing them about the country (Bornemissza, 1960). In the U.K., it is not yet known whether dung-beetles play a significant role in the dispersal of cattle or sheep faeces or whether they could be encouraged.

So here is a small component system that can be studied outside the context of agricultural or even production systems, even though the background argument exists to establish a possible relevance. Questions such as whether dung-beetles significantly influence productivity are secondary to questions about what they do: it is not, of course, being argued here that because the system exists it should have any particular research priority (see Chapter 10 for a discussion of the difficulties of establishing research priorities).

Nevertheless, the meaning attached to "what they do" does relate to the background. For instance, we are interested in whether, or how quickly, or in what conditions, they bury faeces (of what animals) *as a whole*, but traces of faeces may remain in sufficient quantity to influence grazing animals; or the smell alone may be important. So we cannot specify in advance just which aspect of what they do is going to matter, but we do know that quantitative accumulation would be important if no dispersal took place. There are subsidiary questions also, concerned with such matters as the effect of

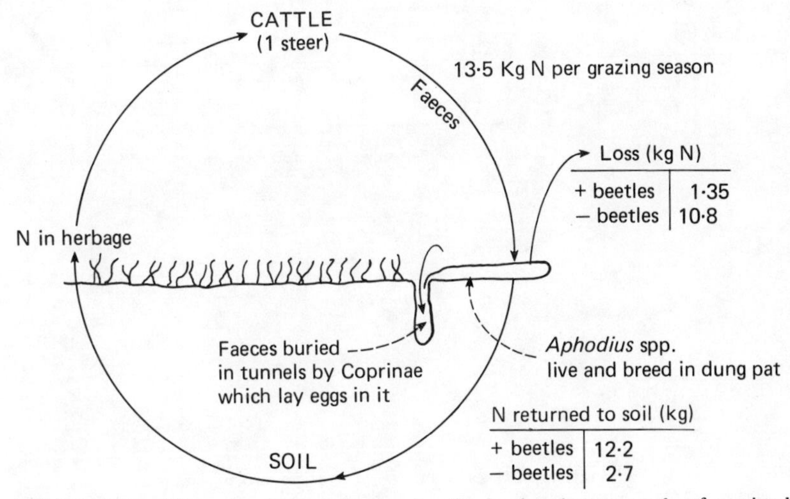

FIGURE 3.10 The role of coprophagous beetles in the nitrogen cycle of grazing land in Australia (based on data from Gillard, 1967).

dung-beetle activity on the eggs and larvae of nematode parasites in the dung (Bryan, 1972).

The component system we wish to understand can be described and pictured without reference to products: only when it is better understood is it possible to convert it, or incorporate it, into a sub-system of relevance to a production system. Fig. 3.10 shows one way in which this component system might be described.

b. Rumen Function

Herbivorous animals are not generally equipped to digest cellulose, a major constituent of their diets, directly, but rely upon micro-organisms that live in symbiotic relationship with them. These micro-organisms are usually concentrated in one part of the alimentary tract. In horses it is the colon, rabbits have an enlarged caecum but also consume their own faeces, and the ruminants (including cattle, sheep, goats, deer, camels and antelopes) are characterized by the possession of an enlarged first part of the stomach, the rumen.

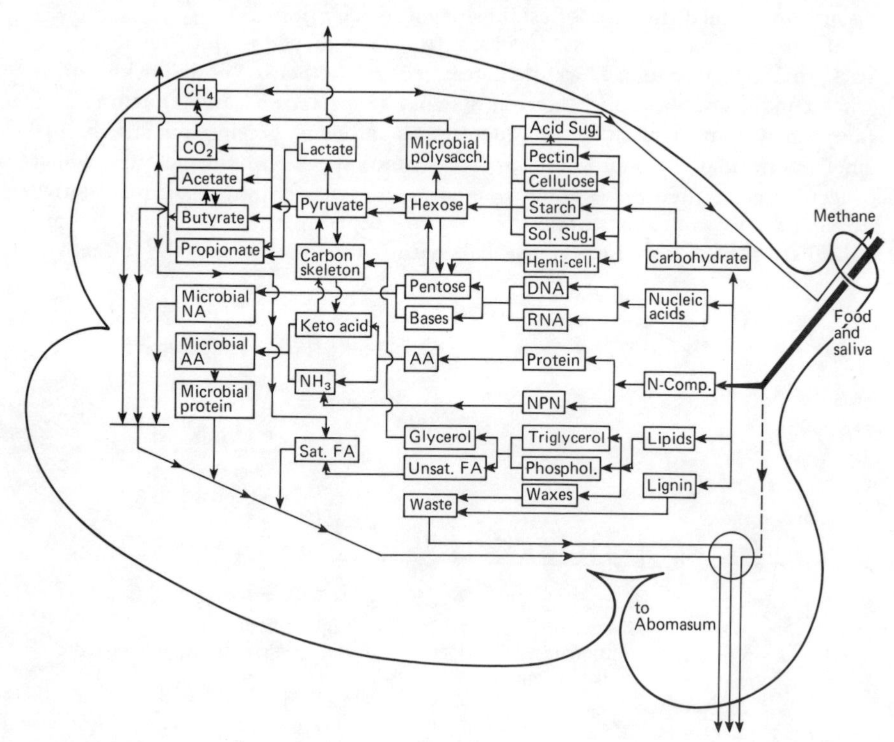

FIGURE 3.11 A diagram of the rumen to show the main biochemical changes taking place within it (Czerkawski, 1974).

Feed flows into the rumen, whence it may also be regurgitated to allow "chewing the cud", and is digested by millions of micro-organisms, largely bacterial flora and unicellular fauna. The proceeds of this digestion may be belched out as methane, absorbed through the rumen wall as volatile fatty acids or pass on to the stomach for further digestion and absorption.

Secretions flow into the rumen directly and, as saliva, in the process of chewing and swallowing. Clearly, the rumen is an important component system, worth understanding even when it cannot be directly related to a product. Figure 3.11 shows one way in which the rumen system may be described.

c. Milk-Fat Synthesis

Milk production is one of the more important production processes, both for direct human consumption and for rearing mammalian progeny. It cannot be completely understood in terms of food going into the animal and milk being secreted: it is worth remembering that it cannot even continue to be secreted if it is not being removed.

The synthesis and secretion of milk by the mammary glands of the udder therefore represent an important component system. But, milk is not one substance, of constant composition, and some constituents may be of greater moment than others. Milk-fat is one of these and its synthesis provides another example (see Fig. 3.12) of a component system.

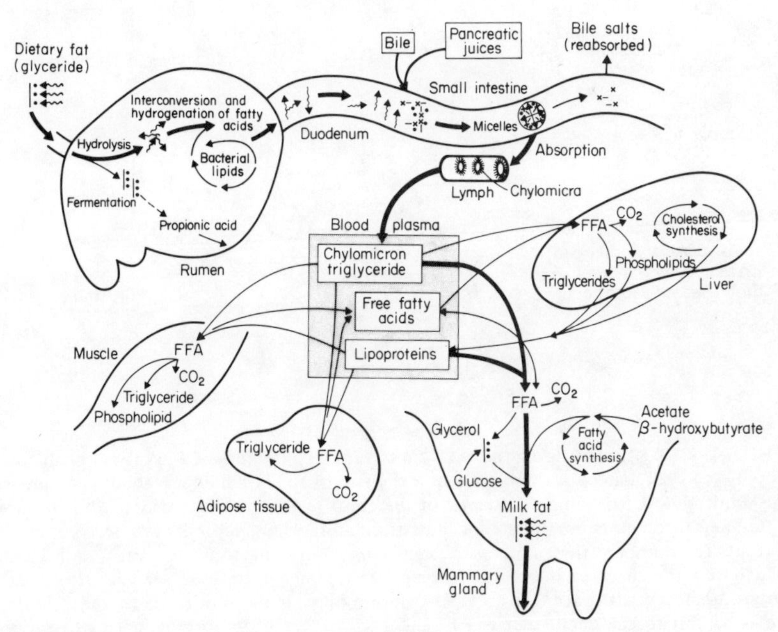

FIGURE 3.12 Digestion and utilization of dietary fat in relation to milk fat secretion (from Storry, 1972).

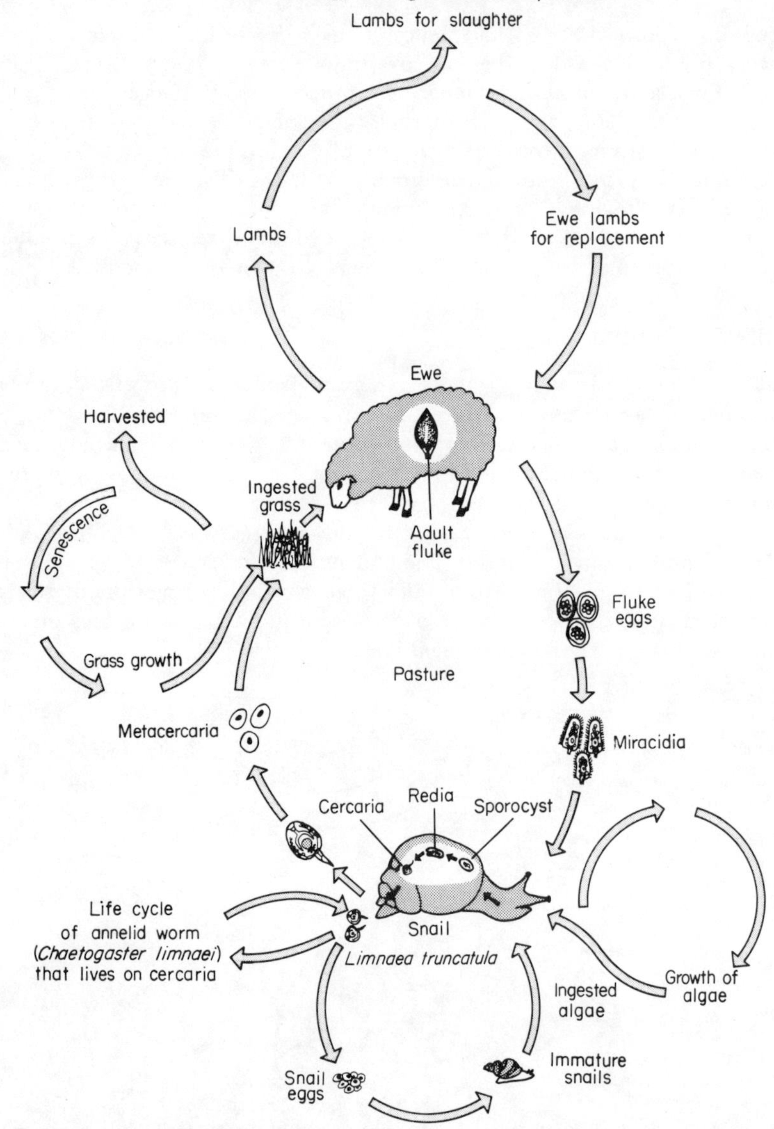

FIGURE 3.13 Life-cycle of the liver fluke (after Pantelouris, 1965). *Note* problems of conveying (a) scale (sheep are approx 1 m × the size of snails, which are about the same size as the adult fluke); (b) time (the rates of different processes differ widely and they vary with temperature, moisture, plane of nutrition and other factors: even seasons are not indicated); (c) number (the numbers of eggs, snails and sheep differ greatly and it is not clear whether the diagram is on a per-sheep basis or per hectare or per year: losses at all stages are unclear); (d) space (the areas or volumes occupied cannot be conveyed); (e) detail (there is a limit to the detail that can be included, either of important factors or physical features: the latter means that the precise point of entry of miracidium to snail cannot be given – yet it matters a great deal).

d. The Liver-Fluke

Many parasites of domestic animals live part of their lives on their host animal and part away from it. The second phase often involves different stages in the life-cycle and serves purposes of dispersal and avoiding too great a build-up of parasites on (or in) one host animal. In the grazing situation, internal parasites frequently spend these phases on the herbage, where they are ingested with the feed. Quite commonly they are voided as eggs in the host faeces and their distribution depends initially on that of the faeces. Survival of stages outside the host may be influenced by the weather, agricultural operations and the activities of other animals.

Some parasites, however, have even more complicated life-cycles and may pass through a secondary host and include phases of further multiplication.

The liver-fluke (*Fasciola hepatica*) is one of these, spending part of its time in sheep or cattle livers and bile ducts, part of its time on herbage and part in a small mud-living snail (*Limnaea truncatula*). Since the parasite can have serious consequences for the host animal and since these depend greatly on the number of parasites, this component-system of the liver-fluke life-cycle is of considerable importance (see Fig. 3.13).

e. Nodulation in Legumes

The main nutritional limitations to the growth of agricultural plants vary with soil nutrient status. Trace elements may be limiting in some areas, potash and phosphate in others; water is limiting in many large areas or at particular times. Amongst the most important, quantitatively, is nitrogen, in spite of the fact that a very high proportion of the aerial environment consists of it.

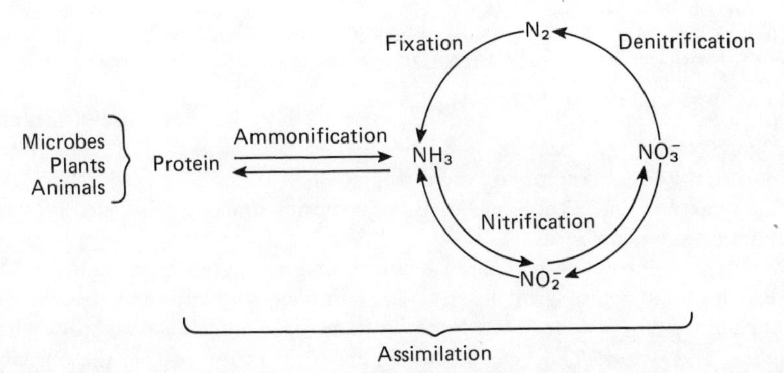

FIGURE 3.14 The biological nitrogen cycle. Plants and microbes make their component proteins from nitrates by way of nitrites and ammonia. Animals make their protein directly or indirectly from plants or microbes. Death, decomposition and putrefaction lead to release of the protein N as NH_3 which may be oxidized back to nitrate by nitrifying bacteria (*Nitrosomonas, Nitrobacter*). Denitrifying bacteria may, where oxygen is not plentiful, reduce the nitrate to N_2, thus causing a net loss of N to the atmosphere; this loss is reversed by the nitrogen-fixing bacteria (from Postgate, 1972).

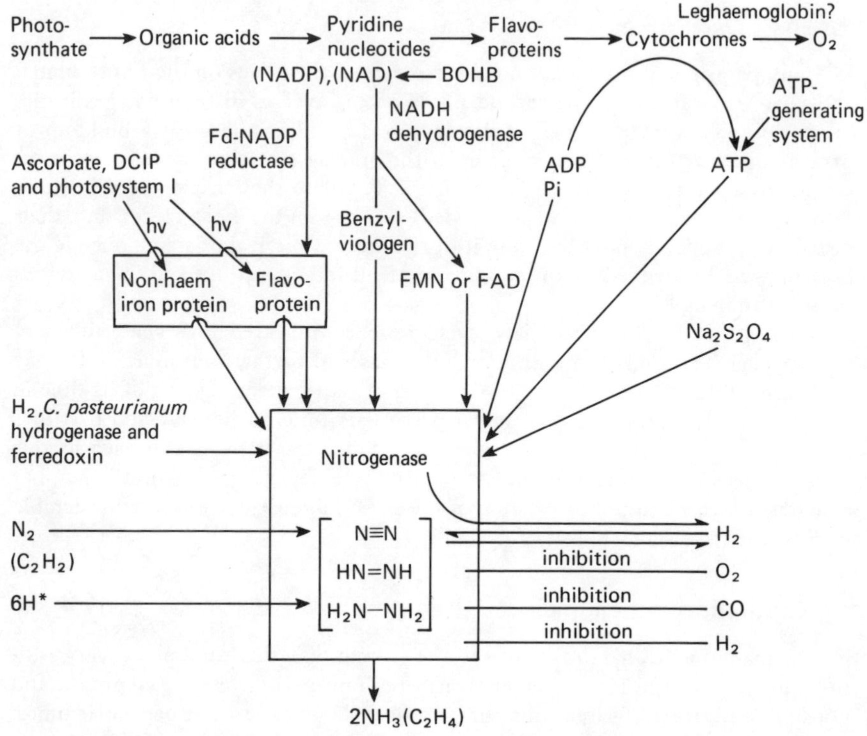

FIGURE 3.15 A schematic representation of N_2 fixation and related processes in legume-root nodules. Reactions demonstrated to occur by use of cell-free systems are indicated by solid lines. An exception is oxidative phosphorylation coupled to electron transport which is based on inhibitor evidence. The arrow from NADP to non-haem iron protein and flavoprotein (both enclosed by a solid line) indicates that both components are essential for electron transport from NADP to nitrogenase (from Evans and Russell, 1971).

Leguminous plants have the advantage that they can use atmospheric nitrogen, fixed by bacteria of the genus *Rhizobium*, and thus operate independently of the supply of soil nitrogen.

The bacteria have to be of the right species and are generally located in nodules on the plant roots.

These component systems are obviously of great agricultural value and they have considerable biological interest as examples of symbiosis between plants. The bacteria supply the host plant with nitrogen and receive carbohydrates, minerals and water. They can live independently in the soil, in the absence of the host plant, and the latter can live independently provided that soil nitrogen is adequate. The essentials of such systems are illustrated in Figs 3.14 and 3.15.

f. Photosynthesis

There is no more important biological process than photosynthesis, the process by which sunlight is absorbed, mainly by chlorophyll in the green parts of plants, and energy is liberated to produce all the organic compounds required

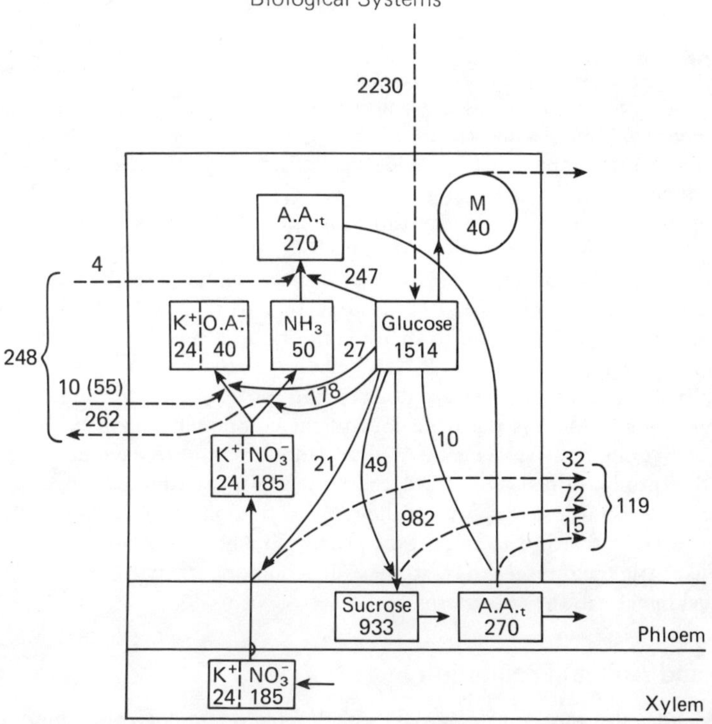

FIGURE 3.16 A simple representation of a photosynthesizing organ: numbers represent wt in g (from Penning de Vries, 1972).

by plants. Yet, within any production system, photosynthesis is only one of a number of component processes. It is thus an example of a component system that is fundamentally similar in a great many production systems: ways of expressing it are therefore of greater significance than would necessarily be the case for many other component systems.

One appropriate level for expressing photosynthesis is biochemical (see San Pietro *et al.*, 1967) and involves considerable complexity. Mathematical models have been constructed both to summarize the operation and results of the process and also to simulate it dynamically (de Wit and Brouwer, 1968). Figure 3.16 shows an example of the latter in flow-diagram form.

These examples have been chosen to cover a range from the obviously important to the superficially trivial, from those that are highly specific to one production system to those of wide generality, including some that operate entirely within an individual and those that involve interactions between plants and animals.

It will be clear that these component systems need to be expressed in greater detail than is possible when dealing with a whole production system: sometimes this is at the same *level* as the system components but sometimes it has to be at a "lower" level (i.e. one involving much greater sub-division into much smaller units). The problems of studying these component systems are eased if the purpose is clear.

Purpose

A production system has a purpose, or perhaps several purposes: they are relatively unambiguous and the purpose of understanding any production system is related to the purposes for which it exists. Clearly, the situation is quite different with component systems; in passing, it may be noted that the purposes of sub-systems are the same as those of the whole system (i.e. in the usage adopted in this book).

The object of understanding a component system is generally wider than just to understand better one part of a system. It is to try and understand how the component works in relation to many contexts and thus the way it would work in many systems. This, however, is a very large task and there is always a risk that it will take a long time and may not provide, in the short term, any information at all about the behaviour of the component in a particular system. Since the component systems themselves cannot be said to have purposes, in the sense that production systems have, the only way to narrow the range of study is to restrict the purposes for which they are to be understood. This can be most readily done in relation to important production systems and, unless it is done, there is little guarantee that studies of components will necessarily be of practical benefit in the short term.

Crop and Animal Production

Agricultural crops and animals are the major biological components of production systems and can be studied as component systems. To study them as isolated components would be less helpful in an agricultural context, although it is very important to do so in parallel with the more applied investigations and such studies certainly fall within the sphere of the biology of agricultural animals and plants.

Crops and animals do not *produce* in isolation, however, but operate primarily as biological systems. These are most easily considered separately, but this is not wholly satisfactory, for example, with grazing systems. However, the major characteristic of a grazing system relates to the typical method by which the animal obtains its food. Where animals are fed by man on feeds that he has harvested, there can be no effect of the animal on crop growth.

In grazing, on the other hand, it is the animal that does the harvesting and, in doing so, has a number of important effects on the plant population.

For this reason, the principles of crop and animal production are dealt with separately (in Chapters 4 and 5) and grazing systems are also included in Chapter 6, which deals with harvesting the product.

This separation of animals and plants for separate discussion is probably sensible and less confusing than trying to deal with them together. Nevertheless, a systems approach to agricultural biology cannot depend upon having special knowledge, only on the ability to obtain or retrieve it. Specialization, including that relating to animals and plants, is thus irrelevant unless they represent totally different kinds of systems. Much has been written on the differences between them and perhaps too little attention has been focused on their similarities.

One interesting way of looking at their differences has arisen from the laboratory need to construct controlled environment chambers for studying them.

Table 3.1 shows the result for a comparison between higher plants and animals.

TABLE 3.1 Principal differences between higher plants and animals (from Housley, 1973)

Character	Higher plant	Higher animal
1. Nutrition	Primary synthesis from simple substances (carbon dioxide, water, salts) via photosynthesis with light as the energy source: complex substances formed.	No primary synthesis: secondary synthesis after digestive breakdown of complex substances from plants or other animals.
2. Body size	Extensive control by external environment.	Limited control by external environment.
3. Body growth	Potentially unlimited in special regions for growth: limitations often set by the external environment.	Development has set proportions irrespective of the external environment.
4. Water relationships	i. Turgor for structural support ii. Transpiration, transport of nutrients. iii. Medium for metabolism.	i. Urinary excretion medium. ii. Body heat regulation in some mammals. iii. Medium for metabolism.
5. Excretion	i. External: respiratory carbon dioxide. ii. Internal: storage of waste products in cell vacuoles.	i. Respiratory carbon dioxide. ii. Soluble products in urine. iii. Solid faecal complex.
6. Nervous system	None.	Present.
Central nervous system	None.	Degrees of intelligence present.
7. Locomotion	None.	Effected by co-ordinated contractile protein in muscle.
8. Body temperature	Unregulated: governed by the environment.	Regulated: independent of the environment.
9. Exchange of respiratory gases	By diffusion.	By forcible exchange: diffusion at cellular level.
10. Biological energy	No reserve pool.	Reserve pool present.
11. Organ number	Very variable, may be high (e.g. leaves).	Low and constant number.

TABLE 3.1—*continued*

Character	Higher plant	Higher animal
12. Structural support	Mainly internal: can have "external" component. i. Turgor of living cells. ii. Thick walls of some living cells. iii. Thickened walls (wood) of dead cells.	Internal vertebrate skeleton.
13. Cell	i. Wall encloses the protoplast. ii. Usually one large vacuole present. iii. Many dead cells may be present. iv. Cell loss is seasonal.	i. Nothing equivalent to plant wall. ii. Usually several small vacuoles present. iii. Few dead cells present. iv. Cell loss is regular compared with plants.
14. Reproduction.		
Asexual	Relatively common compared with animals.	None.
Sexual	Restricted compared with animals: many plants do not bring sexual cells together by their own movements.	Obligatory: sexual cells brought together by animal's own movements.

References

Bornemissza, G. F. (1960). *J. Aust. Inst. agric. Sci.* **26** (1), 54-56.
Brockington, N. R. (1971a). *Span* **14** (1), 1-4.
Brockington, N. R. (1971b). *In* "Grassland Ecology" (C. R. W. Spedding, ed.), pp. 196-204. Oxford University Press.
Bryan, R. P. (1972). *Aust. J. agric. Res.* **24**, 161-8.
Clements, R. O. (1973). *A. Rep. Grassld. Res. Inst.* 1972. p. 131.
Czerkawski, J. (1974). Personal communication.
de Wit, C. T. (1972). Personal communication.
de Wit, C. T. and Brouwer, R. (1968). *Angew. Bot.* **42**, 1.
Evans, H. J. and Russell, S. A. (1971). *In* "The Chemistry and Biochemistry of Nitrogen Fixation" (J. R. Postgate, ed.), Chapter 6. Plenum Press, London.
Gillard, P. (1967). *J. Aust. Inst. agric. Sci.* **33** (1), 30-34.
Hilder, E. J. (1966). Proc. Xth Int. Grassld. Congr., Helsinki, 977-81.
Housley, S. (1973). Lab. Pract., Feb. 1973, 128-132.
Jones, J. G. W. (1970). "The use of models in agricultural and biological research". (J. G. W. Jones, ed.), Proc. Symp., Feb., 1969, Grassland Research Institute, Hurley, Berks.

Lieth, H. (1971). Proc. Brussels Symp., 1969, on Productivity of forest ecosystems. UNESCO, 1971.

Lindemann, R. L. (1942). *Ecology* **23** 399-418.

Macfadyen, A. (1957). "Animal Ecology", Pitman, Lond.

Pantelouris, E. M. (1965). "The Common Liver Fluke", Pergamon Press, Oxford.

Patten, B. C. (ed.) (1971). "Systems Analysis and Simulation in Ecology". Vol. 1. Academic Press, New York and London.

Penning de Vries, F. W. T. (1972). *In* "Crop Processes in Controlled Environments". (A. R. Rees, K. E. Cockshull, D. W. Hand and R. G. Hurd, eds), pp. 327-346. Academic Press, London and New York.

Phillipson, J. (1966). "Ecological Energetics". Edward Arnold, London.

Postgate, J. R. (1972). "Biological Nitrogen Fixation". Merrow.

San Pietro, A., Greer, F. A. and Army, T. J. (1967). "Harvesting the Sun". Academic Press, New York and London.

Slobodkin, L. B. (1959). *Ecology* **40**, 232-243.

Spedding, C. R. W. (1973). Proc. IIIrd Wld. Conf. Anim. Prod., Melbourne 2.1-2.20.

Storry, J. E. (1972). *J. Soc. Dairy Tech.* **25** (1), 40-46.

Walsingham, J. M. (1974). Personal Communication.

Watt, K. E. F. (1966). "Systems Analysis in Ecology". Academic Press, New York and London.

4 The Principles of Crop Production

Plants are cultivated for many different reasons, not all of them concerned with the production of harvestable crops, and not all of the latter are, strictly speaking, agricultural.

The classification of plants* according to the main purposes for which they are grown, although somewhat arbitrary, is a useful way of dealing with the sheer variety of species to be considered.

TABLE 4.1 Major plant families and the crops which are classified in them (after Dittmer, 1972)

Family	Crops
Gramineae	Maize, rice, wheat oat, barley, millet, sorghum, sugar cane, and numerous other grasses
Leguminosae	Bean (lima, kidney, etc.), pea, peanut, soybean, alfalfa, clover, lentil, cowpea, tamarind
Rosaceae	Apple, pear, plum, cherry, peach, almond, apricot, raspberry, strawberry, blackberry, nectarine, loquat, quince, and others
Vitaceae	Grape
Ericaceae	Blueberry, cranberry, and huckleberry
Rutaceae	Orange and all other citrus fruits
Moraceae	Mulberry, breadfruit, and fig
Bromeliaceae	Pineapple
Musaceae	Banana
Euphorbiaceae	Cassava or manioc (tapioca)
Chenopodiaceae	Beet, spinach, Swiss chard, and other leafy vegetables
Umbelliferae	Carrot, celery, parsnip, and parsley
Cruciferae	Cabbage, mustard, radish, turnip, rutabaga, kohlrabi, and others
Solanaceae	White potato, tomato, green pepper, egg plant, and others.
Cucurbitaceae	Cucumber and melon
Palmaceae	Coconut, oil palm and date
Dioscoreaceae	Yam

* See Dittmer (1972), Hutchinson (1969), Kipps (1970) and Schery (1972).

One way of describing the main purposes is as follows:

1. To clothe land for amenity reasons;
2. To produce crop products for direct consumption or use;
3. To produce animal feeds;
4. To perform a function *in situ* (e.g. shelter belts or erosion control);
5. To increase soil fertility;
6. To utilize environments that impose serious limitations on what can be grown. (This will tend to include one or more of the foregoing purposes.)

However, whilst it is useful to identify these purposes there does not seem to be great value in listing all the plants involved under these headings. In fact, since plants of the same family often produce a similar product, there is much to be said for considering the plant species that are used for agricultural purposes, within the main families that have been cultivated for food (Table 4.1). Plants that have been specially used for the making of beverages (Table 4.2), oils, waxes, pectins, gum and resins (Table 4.3) or perfumes and flavours (Table 4.4), cover a wide range of species.

Plants have many other uses as well: vegetable tannins come from oak, hemlock, the South American quebrancho tree and mangroves; dyes come from such plants as *Indigofera tinctoria* (for indigo); medicines from fruits, such as

TABLE 4.2 Beverage plants (after Dittmer, 1972)

Coffee	*Coffea arabica* (Rubiaceae)
Tea	*Thea sinensis* (Theaceae)
Maté	*Ilex paraguayensis* (Aquifoliaceae)
Chocolate and cocoa	*Theobroma cacao* (Sterculiaceae)
Alcohol	Fermented grains, roots, tree sap and fruits
Wines	Fermented grapes

TABLE 4.3 Vegetable oils and other constituents (After Dittmer, 1972)

Vegetable oils	come from cottonseed maize oil palm olive peanut safflower soybean (also castor bean, almond, sunflower seed, linseed)
Waxes	from plants. exudate from leaves of Brazilian palm (*Copernicia cerifera*)
Pectins	from citrus and green apple residues
Gums	many from legumes
Resins	mostly from Pinaceae

TABLE 4.4 Perfumes and flavours (After Dittmer, 1972)

Family	Plant name	Product
Labiatae	*Mentha piperata*	peppermint
	Melissa officinalis	balm
	Salvia officinale	sage
Rosaceae	*Prunus amygdalus*	oil of almond
Compositae	*Artemisia dracunculus*	tarragon
Orchidaceae	*Vanilla* spp.	vanilla
Zingiberaceae	*Zingiber officinale*	ginger
Liliaceae	*Allium sativa*	garlic
	Allium cepa	onion
	Smilax spp.	sarsaparilla
Cruciferae	*Rorippa armoracia*	horseradish
	Brassica alba	yellow mustard
	Brassica nigra	black mustard
Lauraceae	*Cinnamomum zeylanicum*	cinnamon
Myristicaceae	*Myristica fragrans*	nutmeg
Myrtaceae	*Eugenia caryophyllata*	cloves
	Pimenta officinale	allspice
	Eucalyptus spp.	eucalyptus oil
Moraceae	*Humulus lupulus*	hops
Umbelliferae	*Apium graveolens*	celery
	Coriander sativum	coriander
	Angelica archangelica	angelica oil
	Foeniculum vulgare	fennel
	Anethum graveolens	dill
	Apium sativum	parsley
	Pimpinella anisum	anise
	Carum carvi	caraway

TABLE 4.5 Fibres (After Dittmer, 1972)

Soft	mainly from phloem
	Flax (*Linum usitatissimum*)
	jute (*Abutilon avicennae*)
	hemp
	ramie (*Boehmeria nivea*)
Hard	from entire vascular bundles
	Manila hemp (*Musa textilis*)
	Sisal (*Agave sisalana*)
Surface fibre	Cotton (*Gossypium* spp.)
	Kapok (Bombacaceae)
	Coir (Coconut palm musk *Cocos nucifera*)

opium from *Papaver somniferum*, and from bark, as in the case of quinine from *Cinchona* spp.; insecticides like roterone (from *Derris elliptica*) and pyrethrum (from *Chrysanthemum coccineum*).

Fibre production is based on plants like flax, jute, hemp, cotton and kapok (see Table 4.5). Some of the foregoing agricultural products come from trees, and the distinction between agriculture and forestry, although both are large subjects, is not helpful if drawn too rigidly in the context of this discussion. A typically borderline product is latex, derived from *Hevea brasiliensis*. It is a short step from the production of agricultural fibres to the production of forest timber, especially when a great deal of this goes to the production of pulp for papermaking.

TABLE 4.6 Hardwoods (After Dittmer, 1972)

1. Oaks (white, red, or black) cabinet work, veneers, tight cooperage, flooring, trim, furniture, and general construction
2. Maples flooring, furniture, musical instruments, veneers, implement handles,. maple syrup.
3. Walnut cabinet wood, gunstocks, edible nuts
4. Beech furniture, flooring, woodenware novelties
5. Hickories axe handles, wheel spokes, hickory and pecan nuts
6. Cherry furniture, veneers, handles, fruit
7. Ash tool handles, baseball bats, oars
8. Sycamore butcher's blocks, musical instruments; especially good when quarter-sawed for veneers, trim, furniture
9. Gums cheap veneers, veneer cores, wood rollers, gunstocks
10. Poplars pulp, boxes, crates, excelsior, fuel.

TABLE 4.7 Softwoods (After Dittmer, 1972)

1. Douglas fir lumber and general heavy construction, plywood; highest quality softwood in the U.S.
2. Redwood shingles, timbers, tanks, tight cooperage, caskets
3. Hemlocks source of tannin, boxes, pulp, cheap construction
4. Spruces paddles and oars, pulpwoods, interior trim, ornamental planting
5. First (white and balsam) paper pulp, boxes, crates, Christmas trees, resinous blisters on bark form Canada balsam
6. Arborvitaes shingles, boats, cooperage
7. Pines white used for many years for ship masts
   Red pine crates, pulp
   Long-leaf pine general construction, boxes, crates
   Western yellow pine general construction, boxes, crates
   Lodgepole pine posts, telephone poles, construction lumber
   Pinon pine edible seeds, fuel
8. Cedar cedar chests, closet linings, pencils, woodenware

TABLE 4.8 Proportion of calorie intake provided by different food categories (after Brown and Finsterbusch 1972, based on USDA data)

Major Crops	%
Rice	21.2
Wheat	19.6
Maize	5.4
Potatoes	4.9
Millet and sorghum	4.1
Cassava	2.0
Rye	1.6
Barley	1.5
All other foods	39.7

These, then, are the main plant species and their products: their relative contribution to man's diet is indicated in Tables 4.8 and 4.9, in terms of the proportion of calorie intake provided by different crops. The relative agricultural importance of the major crops is illustrated by the proportion of the total cultivable area devoted to them (Table 4.10).

An excellent idea of the range of products based on wood can be gained from the information given by Dittmer (1972), in his book on "Modern Plant Biology" (see Tables 4.6 and 4.7).

TABLE 4.9 Calorie Intake (%) (after Brown and Finsterbusch, 1972)

	Grain products roots and tubers	Fruits, nuts and vegetables	Sugar	Fats and oils	Livestock products	Fish
World average	62.7	9.6	7.3	8.9	10.8	0.7
Regional variation (geographical)	24.4–74.5	3.5–12.3	4.1–16.3	5.3–19.9	3.8–35.2	0.2–0.9
Economically developed regions	47.3	5.9	11.1	14.5	20.7	0.5
Less economically developed regions	71.7	11.5	5.1	5.8	5.1	0.8

TABLE 4.10 World harvested area of principal crops
(Source: Brown and Finsterbusch, 1972, based on USDA data)

Crop	% of total area
Wheat	22.1
Rice	12.7
Corn	11.4
Millet and sorghum	10.1
Barley	6.6
Oats	5.0
Rye	3.3
Total grains	71.2
Oilseeds	7.2
Roots and tubers	5.0
Pulses	4.9
Fibres	4.7
Fruits and vegetables	3.7
Sugar	1.5
Beverage crops	1.0
Rubber	0.4
Tobacco	0.4
	100.0

TABLE 4.11 Net primary productivity of major vegetation units (after Lieth, 1972)

	Area (10^6 km^2)	Total net primary productivity for area (10^9 t)
Continental		
Forest	50.0	64.5
Woodland	7.0	4.2
Dwarf and open scrub	26.0	2.4
Grassland	24.0	15.0
Desert	24.0	–
Cultivated land	14.0	9.1
Fresh water	4.0	5.0
Total	149.0	100.2
Oceanic		
Reefs and estuaries	2.0	4.0
Continental shelf	26.0	9.3
Open ocean	332.0	41.5
Upwelling zones	0.4	0.2
Total	361.0	55.0
Total earth	510.0	155.2

The relative primary productivity of the major vegetational types has been estimated by Lieth (1972) for the year 1950 (see Table 4.11).

The amounts, rates and patterns of production for these areas are dominated by the environment.

The Crop Environment

Many components of the environment influence the plant species that can be grown, their productivity and their costs of production.

Light, temperature and water supply are of over-riding importance, but the supply of mineral nutrients is of great agricultural importance.

Sunlight

The energy source for green plants is solar radiation in the 0·4–0·7 μm wave region of the spectrum. Light intensity is important, since species vary in the extent to which they can make use of different light intensities. Daylength is also of great consequence and both daylength and light intensity are influenced by latitude, altitude and such features as cloudiness.

All points on the earth receive the same amount of daylight per year (MacArthur, 1972): disregarding light intensity and the effects of cloud, dust and vegetational cover, each point receives in total, just over half-a-year of day-

TABLE 4.12 Regional variation in maximum crop growth rate (after Cooper, 1970)

Climate and country	Crop	Crop growth rate ($g\ m^{-2}\ day^{-1}$)	Total radiation input ($cal\ cm^{-2}\ day^{-1}$)	Conversion of light energy (%)
Temperate				
U.K.	*Lolium perenne*	16·6	290	5·4
U.K.	Barley	23	484	4·0
U.K.	Beet	31	294	9·5
New Zealand	*L. perenne*	17·5	480	3·4
U.S.A.	*Zea mays*	52	500	9·8
Subtropical				
U.S.A.	*Zea mays*	52	736	6·4
Australia	*Digitaria decumbens*	19·3	580	3·1
Tropical				
Puerto Rico	*Panicum maximum*	16·8	480	3·3
Australia	*Pennisetum typhoides*	54	510	9·5

light each year. The distribution and pattern of daylength and the light intensity are, of course, very different at different sites.

Growing plants use two main sources of light, direct sunlight and diffuse skylight, and the green part of the plant may be regarded as a light-collecting surface. Since the sun moves continuously, effective light collection is not a simple matter.

Watson (1947) introduced the concept of leaf area index (L) to denote the total area of leaf per unit area of land. This gives an indication of the area available for photosynthesis, although stems, petioles, leaf sheaths and inflorescences also contribute to photosynthesis in many plants.

The amount of leaf area is clearly important, as is leaf angle and shape: indeed the "geometry" of the crop (i.e. its distribution in space) is of great

TABLE 4.13 Maximum growth rates of selected crop species (after Cooper, 1970)

Climatic region	Crop species	Crop growth rate ($g\ m^{-2}\ day^{-1}$)	Conversion of light energy (%)
Temperate	Perennial ryegrass	16·6	5·4
	Barley	23·0	4·0
	Beet	31·0	9·5
	Potato	22·8	5·4
	Maize (N.Z.)	29·2	6·1
	Maize (U.S.A.)	52·0	9·8
Subtropical	Sorghum	51·0	6·7
	Maize	52·0	6·4
	Cynodon dactylon	21·2	3·8
Tropical	*Pennisetum typhoides*	54·0	9·5
	Digitaria decumbens	26·4	5·8

importance, especially in a competitive situation. The latter is in fact, the normal one, since competition may be between individuals of the same species as well as with other plant species. Mutual shading between plants, and between leaves on the same plant, may occur in foliage dense enough to collect a high proportion of the incident light. This is only disadvantageous if the light intensity is then too low for any one leaf.

Plants vary in their capacity to photosynthesize at low light intensities and in the maximum rates they can achieve (see Tables 4.12 and 4.13) but the maximum for individual leaves of agriculturally important species was estimated by de Wit (1967) as about 20 kg carbohydrate ha^{-1} h^{-1}.

Crops also differ in the leaf area index values they can sustain and the duration of this leaf area in terms of the period of the year for which it lasts.

Temperature

This depends on latitude, and varies seasonally and from day to night: it also varies with altitude, becoming lower by about 10°C (5·5°F/1000 ft) for each km of height under dry conditions (MacArthur, 1972). This "adiabatic lapse rate" is reduced under humid conditions to about 6°C/km (3°F/1000 ft).

The response of plant species to temperature changes varies enormously but plant growth is generally slow below 7°C (45°F) and few plants are adapted to temperatures much in excess of 38°C (100°F).

Water Supply

The total quantity of water in the world is enormous, but some 97% is estimated to be in the oceans; about 2% is assessed as frozen and only 0·8% regarded as "available" (Dittmer, 1972). Most areas of the world depend upon precipitation, which varies from virtually zero to several hundreds of inches per year but both annual and seasonal variations are considerable. Evaporation rates also vary greatly but in the tropics rates rarely exceed 55 in (= 1397 mm) per year (and, in vast areas, do not exceed 45 in (= 1143 mm) per year). Since rainfall may be many times this amount, the run-off is considerable (Hoare, 1967). Tropical rivers, such as the Nile and the Indus, thus collect enormous quantities of water that can be used for the irrigation of arid areas. Snow melt also feeds rivers on a large scale (e.g. the Ganges, the Tigris and Murray rivers).

Some of these vast rivers are often destructive (e.g. the Darmodar in India and the Ord in Australia) and others (Amazon and Orinoco in South America; Niger and Zambezi in Africa; Irrawaddy and Mekong in South East Asia; Yangtze and Yellow river in China; Ob and Lena in U.S.S.R.) pass through lands that are not easily developed for irrigation. Where water is available, there are many methods of making use of it for irrigation purposes (see Table 4.14).

Light, temperature and water supply can all be manipulated but, in practical terms, the first two only to a very limited extent. Water supply can be augmented by irrigation on quite a large scale (Hoare, 1967), but from a world point of view, all three factors may only be controlled locally. Modifying the

TABLE 4.14 Methods of irrigation (After Ruthenberg, 1971)

Water Supply	Diversion of natural flows (e.g. rivers) Damming (of tidal, river or rain-water) Pumping (by wind-, water-, animal-, man- or motor-power)
Water distribution	Ditches Flumes Pipelines
Water application	Flooding Basin irrigation Furrow irrigation Sprinkler irrigation Underground infiltration

environment for crop growth is generally very expensive, although for high-value, horticultural crops, for example, very extensive areas may be covered with glass or polythene and may be supplied with additional heat, water and nutrients.

Full irrigation, maintaining the soil at or near field capacity, results in the highest crop production but does not necessarily increase the yield of product proportionately. This is especially so for non-foliar products. For example, when irrigation increases the yield of broad bean (*Vicia faba*) foliage by 50%, the marketable weight may only rise by 8% (Winter, 1967): comparable increases for carrots were a 100% foliage increase with only 20% increase in marketable yield. Furthermore, in many crops there are moisture-sensitive growth stages when irrigation can have spectacular effects on yield. In the pea crop, increases in yield of peas of up to 60 cwt/acre were obtained by watering at the start of flowering (to increase the number of peas per pod) and when the pods were swelling (to increase the size of each pea).

Where the environment cannot be greatly changed, it is necessary to choose suitable crop species to grow. Since both water and mineral nutrients are mainly derived from the soil, the capacity of the plant's root system to explore and exploit the soil should be borne in mind in the selection of species (Scott Russell, 1971).

Few of these species could sustain high yields, however, without additional nutrients.

Fertilizers

The reason for this is precisely because the choice of species is based on the supply of water, light and heat in the environment, and not on the supply of nutrients (with some notable exceptions—see below). Nor would there generally be much point in choosing a crop species that could perform at a low level of fertility, since this may also imply low crop yields, where fertility could be improved by the addition of fertilizers. In the very nature of cropping, nutrients are removed in the harvested material and subsequent yields depend upon their replacement. Since the environments chosen usually provide ample gaseous CO_2 and N and adequate water and light, the part of the harvested crop that clearly requires replacement is the mineral content. Some of these minerals are only required in small amounts and can be supplied by the soil: many important ones such as phosphorus may be involved in complex nutrient cycles that influence availability to plants (Floate, 1970).

The position of nitrogen is exceptional, however. It is an astonishing fact that most grasslands, for example, would produce vastly more herbage if they were supplied with additional nitrogen, yet they grow at the interface of an atmosphere that consists of 80% N (representing some $3{\cdot}7 \times 10^{15}$ tons of N in total: McKee, 1962) and a soil that often contains a total of 2000 to 5000 kg/ha N, in the top 6 in of English soils, for example (Cooke, 1967). The sedimentary rocks of the earth are estimated to contain about 4×10^{14} tons, the seas 2×10^{13} tons and soils about 15×10^{10} tons; by contrast plants contain a total of only 1×10^{9} tons and animals 6×10^{7} tons (Donald, 1960).

The problem, of course, is that the rate of incorporation in plants from either source, air or soil, is relatively slow. Some plants depend entirely on nitrate supplied in soil water. Others, notably the legumes but also such plants as the blue-green algae that sometimes influence rice production, are able to fix atmospheric nitrogen (see Russell, 1961). In the case of the legumes, this is accomplished by bacteria living in nodules on the roots of the plant. They not only supply the legume crop with nitrogen but also supply other plants indirectly when legume leaves and roots die and are incorporated in the soil.

Some plant groups can absorb ammonia and the number able to fix atmospheric nitrogen may be much larger than is currently supposed (Nutman, 1972).

However, in general, where non-leguminous agricultural crops are grown the yield level is usually heavily dependent on the supply of nitrogenous fertilizer. The quantity used has increased steadily from a total of some $3{\cdot}5 \times 10^{6}$ tons (N) applied between 1837 and 1939, to about $1{\cdot}6 \times 10^{6}$ tons from 1940 to 1949 and $1{\cdot}9 \times 10^{6}$ tons (N) from 1950 to 1959 in the U.K. alone (Cooke, 1967). Similarly, vast increases have occurred in the use of potassium (see also Watson, 1970). World consumption of fertilizers (in terms of the total amount of N, P_2O_5 and K_2O) was 48·3 m. tons in 1966/67 (Brown and Finsterbusch, 1972). The consumption per hectare of arable land was 36 kg (mean) but varied from 9 kg in the developing countries to 350 kg in Japan.

Other Aspects of the Environment

There are many aspects of the environment that are important for crop growth, but the major physical features of soil, slope, topography, altitude and geomorphology mostly operate through one or more of the categories already mentioned.

Naturally, these issues have been oversimplified, in order to condense them into a short space: water supply, for example, has been discussed as if the problem was always one of shortage, whereas drainage may often be what is required. Whatever the environment, however, and whatever the product, crop production has to be based on essentially the same crop growth and development processes.

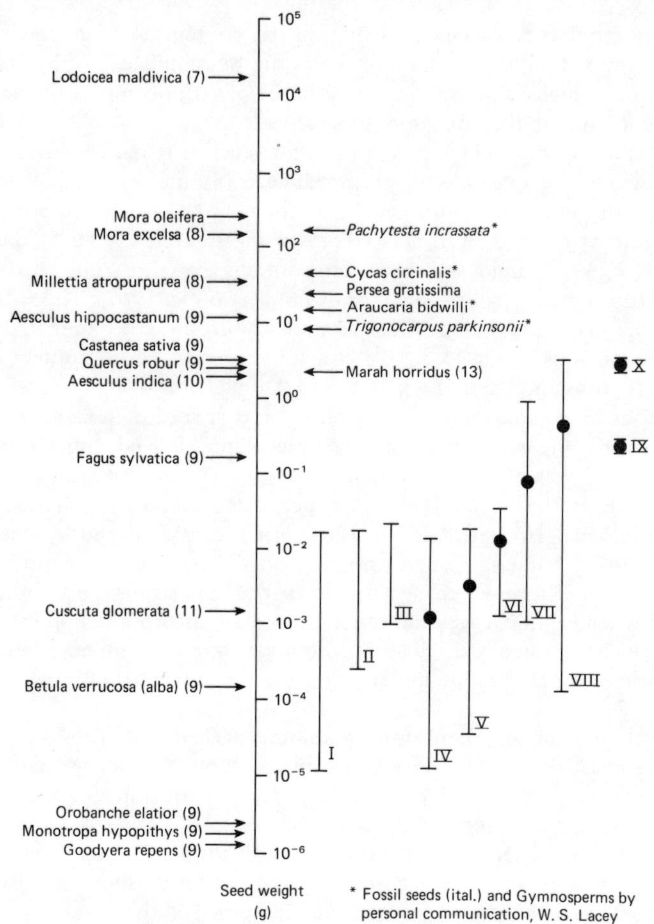

FIGURE 4.1 Ranges of variation in seed weights (from Harper, Lovell and Moore, 1970).

Seed weight ranges of N. American herbaceous plants [Stevens (1932; 1957)].

I Rosaceae, Scrophulariaceae, Compositae, Ranunculaceae, Cruciferae, Gramineae, Cyperaceae, Chenopodiaceae, Caryophyllaceae
II Liliaceae, Polygonaceae
III Leguminosae, Umbelliferae

Seed weights (mean and range) of groups of species from various habitats in Britain, Salisbury (1942).

IV Open habitats, short grass, and meadows
V Woodland margins
VI Woodland ground flora
VII Woodland shrubs
VIII Woodland trees

Seed weights (mean ± S.E.) of Central American woody legumes, Janzen (1969).

IX Species subject to bruchid attack
X Species not attacked by bruchid beetle larvae.

Reproduced, with permission, from "The Shapes and Sizes of Seeds", *Annual Review of Ecology and Systematics*, Vol. 1, pp. 329 and 331.

Crop Production Processing

Establishment of the seed (or its equivalent) is the starting point in the growth of a new individual plant. The variety of seeds and their associated structures is enormous and there are few environmental niches that are not available to one or more plant species. Seeds germinate on walls, under pavements, on rooftops and high up on the branches of trees: but agricultural plants are mostly grown in soil, although some are grown *under* water and some *in* running water (rivers or in flow-culture solutions).

Plants have developed this variety of seed structures in response to the range of environmental opportunities and in relation to the hazards represented by destruction by animals and loss due to other causes. There is, therefore, considerable variation in the number as well as the size and structure of seeds (see Fig. 4.1) and thus in the proportion of plant production devoted to reproduction. The latter has been studied in populations of golden rod (*Solidago* spp.) by Gadgil and Solbrig (1972): some of their results are illustrated in Fig. 4.2. Harper *et al.* (1970) have described the production of annual net assimilation involved in the reproductive effort of several different types of plant (Fig. 4.3).

In natural situations, there is clearly a balance of advantage and disadvantage to be achieved in the proportion of total production that should be directly used

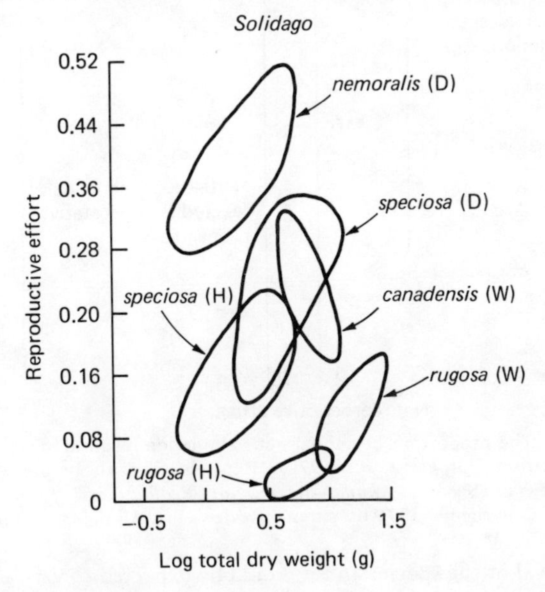

FIGURE 4.2 Reproductive effort (i.e. ratio of dry weight of reproductive to total aerial tissue) as a function of the total dry weight of the aerial tissue for the various populations of *Solidago*. Each closed curve embraces all points representing the individuals included in a single population: D = dry-site population; W = wet-site population; H = hardwood-site population (from Gadgil and Solbrig, 1972; reproduced with permission of the University of Chicago).

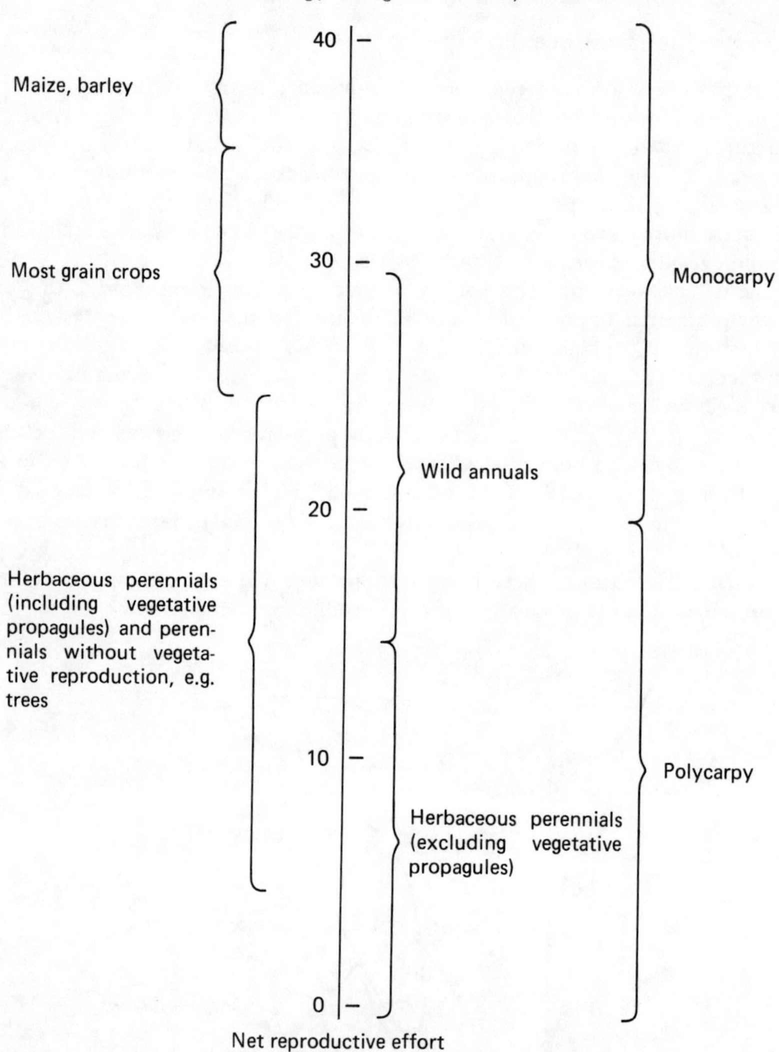

FIGURE 4.3 The proportion of annual net assimilation involved in reproductive effort, Ogden (1968) (from Harper *et al.*, 1970). Reproduced with permission, from "The Shapes and Sizes of Seeds", *Annual Review of Ecology and Systematics*, Vol. 1, pp. 329 and 331. Copyright © 1970 by Annual Reviews Inc. All rights reserved.

to ensure survival of the species and it would be expected to vary with whether a plant is annual or perennial and, in the latter case, with longevity. Furthermore, not all reproduction is by seed: vegetative reproduction is practised in a number of agricultural crops (see Table 4.15).

The "cost" of sowing or planting a new crop may be a major one in agriculture and the proportion of a crop that has to be devoted to ensure

TABLE 4.15 Examples of agricultural crops that are reproduced vegetatively

Crop plant		Organ involved
Potato	*Solanum tuberosum*	Tuber
Cassava	*Manihot esculenta*	Shoot
Yam	*Dioscorea alata*	Tuber
Banana	*Musa* spp.	Shoot (sucker)
Strawberry	*Fragaria* spp.	Rooted runner
Sweet potato	*Ipomoea batatas*	Shoot
Sugar cane	*Saccharum officinarum*	Shoot
Pineapple	*Ananas comosus*	Shoot or crown of leaves on fruit

establishment of the next crop may be important. The cost may be important whether the method of establishment involves the product (as in grains and potatoes) or not (as in turnips or sugar beet).

Growth and Development

Since "growth" can mean increase in size, weight or number, and is often accompanied by differentiation of the growing cells and tissues, it is convenient to refer to "growth and development" together. The whole growth cycle of a plant includes senescence and death; it therefore includes phases of negative growth, represented by weight loss and reduction in size.

When terms are used as widely as this, it is tempting to argue that they should be more precisely defined, but some terms are best kept for this "umbrella" role, covering a range of possible, but related, meanings, provided that, whenever they are used, a precise meaning is attached and stated. Indeed, there are serious dangers of any definition or meaning that is convenient in a particular context gradually being accepted as *the* definition of growth.

Dry matter is a good example in plants. It is used to remove variation between plant weights due to variations in, for example, internal and external moisture. Clearly, the weight of an externally wet plant is a poor index of its real weight or size and changes in such weight may represent a poor index of growth. The usual expression of dry matter, however, not only removes this variation but also the water content of the plants and any genuine variation in that. Plants are also "ashed" to determine total mineral content, but it is unlikely that changes in the weight of total minerals would be described as growth. Changes in dry matter are often used to describe growth, however, and growth is sometimes defined in this way.

Now, of course it is legitimate to be interested in changes in the amount of dry matter present, especially if this is a good way of describing the product, but it is doubtful whether it can generally (or usefully) be described as growth.

In the case of forage species (Arnott *et al.*, 1974), herbivorous animals eat the fresh, wet material and some sap-sucking insects live on the fluids, but few animals actually consume dry matter as such. In plant growth terms, the living processes that constitute growth or on which growth depends can only occur in tissues containing water.

In short, the accumulation of dry matter (or calcium or nitrogen) is one way of looking at growth, or the results of growth, but growth itself can only relate to living tissue containing some water.

In forestry, especially over substantial periods of time, growth may be expressed as increases in the air-dry weight of wood or the accumulation of carbon, on the grounds that, although leaves are grown, they are ephemeral and do not contribute to the total accumulation. Herein lies the key to this discussion, that if something does not contribute to what we are interested in, we tend to leave it out. In other words, we are really talking about *production*, of one sort or another, and the growth of the product; but this is only a part of total growth and it may be better to talk about measuring changes in constituents due to growth, rather than equate such changes with growth itself. In the case of "biological production", however, growth may mean the same thing: it is where the yield relates to only a part of the plant that the difficulty arises.

Similar considerations apply to animal growth (see Chapter 5), but the problems are somewhat different.

In most crop plants, growth occurs, as a result of photosynthesis, provided that the supply of nutrients is adequate and the environment is suitable, in terms of light and temperature.

It may seem obvious that growth is important in crop production but it has to be remembered that growth is not the same as agricultural production. In many cases, the product forms only a small part of the plant and the quantity of product is not always related to total plant growth.

"Yield", or the amount of harvestable material, is the important characteristic, therefore: if it is distinguished from "production" (even agricultural production) it is simply on the basis that not all that is produced can be harvested. This may be because it is too expensive, or impracticable with the means available, or because there are unavoidable losses (see Chapter 7), or because the quality of product that can be marketed is very tightly defined (so that, for example, only unblemished apples of a certain size contribute to yield, out of a much larger, total apple crop).

The most important aspects of growth to the agriculturalist, therefore, are the rate at which it proceeds, the total growth achieved (e.g. annually) and the proportion of this total that is devoted to the product.

It must not be supposed, however, that the aim should be simply to maximize total growth and the proportion going to production. Crop production is usually a much more complex matter. For root crops or tubers, such as the potato, the leaves are vital but it does not follow that more or larger leaves necessarily mean larger roots or tubers.

Wheat and barley production are good examples where progress has been made not only by increasing the total growth of the plant but by selecting varieties that put a higher proportion of their growth into the grain. They are also examples of crops where the potential production is determined quite early in the season: no amount of growth later on can result in production greater than this potential.

Thus, the number of "production-points" may be laid down at particular times and the environment or damage at these times may then be critical.

The yields of fruit trees can be irreparably damaged in the bud stage, of grain crops at tillering or flowering, of grasses at times of tillering, and of potatoes at tuber initiation.

Production is not maximized by *maximizing* the number of buds, seeds, tubers or growing points, however, since these may then be overcrowded and compete for an inadequate total supply of photosynthate (Scott, 1972). Equally, the greatest total yield may not be associated with maximum individual product size. Optimum combinations of both attributes are required.

A very high growth rate may be desirable in the production of non-fibrous edible products; it may be undesirable where the quality of the product may be adversely affected, as with tobacco. Generally, high growth rates are needed in order to achieve potential production within the growing season but it is the production that matters, rather than the *rate* at which it is produced.

There are times when the ability of a plant to grow rapidly is an advantage, in order to exploit scarce or transient resources; on the other hand, it is the capacity of some plants to remain virtually static that enables them to survive very cold or very dry weather.

Maximum growth rates differ substantially between species (see Table 4.13) and may be costly to obtain. When it is argued that a high growth rate is desirable, it is usually assumed that this is within a context of restraints on the use of expensive resources: the heating of the soil of grassland in winter is not envisaged, for example.

Reproduction

Reproduction is of obvious importance in maintaining the species, and the potential rate of reproduction clearly has a large influence on the rate at which plant populations can be multiplied. It is of importance, therefore, in relation to genetic improvement and the speed with which it can be applied in practice.

Furthermore, many important agricultural products are the "fruits" of reproduction and the reproductive process is then also the main production process. Even where reproduction is relatively unimportant, however, it may exert effects on other processes, such as growth, that are directly involved in production. Grasses are an example of this, especially those that can also be propagated vegetatively, since the nutritional value of the main product (the above-ground foliage) may be greatly influenced by the reproductive phases of growth.

Where products are the direct results of reproduction, it is advantageous to have a plant that grows vigorously and devotes a high proportion of its assimilates to reproduction. A similar proposition emerges when considering meat production from animals (see Chapter 5) and it is interesting to consider what the effect of evolutionary pressures has been on this.

As Harper (1967) has pointed out, for a given plant species in a given environment, there must be an optimum partition of assimilates between growth and survival of the individual and reproduction to ensure survival of the species. Obviously, the resources that have to be devoted to the production of spores, seeds and pollen, in terms of their numbers, sizes and content of food reserves, depend upon the hazards that such organisms face between liberation from the parent plant and the establishment of a new individual. Of course, losses may also occur before "liberation" and after establishment but before reproduction of the new individual: a complete calculation has generally, therefore, to take into account at least one generation. Longevity also needs to be taken into account since this influences the total number of progeny produced by an individual in its lifetime.

The range of devices for ensuring successful survival of the species is enormous and there are clearly many different ways of dealing with substantially similar hazards. It is difficult to make sensible comparisons between plant species, although such comparisons would seem to be of great interest. Indeed, there is considerable interest in the comparison of animals and plants in this context.

The problem may be simply stated as a series of questions:

Do some plant species have to devote a much higher proportion of their total assimilates to reproduction? (This has been discussed, in relation to seed size and shape, by Harper *et al.*, 1970).

Is such increased energy (and other) expenditure necessitated by greater hazards or by greater vulnerability (or by a short life-span)?

Are there compensations, presumably involving energetic economies, associated with the greater vulnerability or with the environments that include greater hazards?

Is there some general principle involved, such that, for example, the proportion of energy devoted to reproduction is constant in relation to some expression of the hazards, on a life-time basis?

What are the differences between plants, warm-blooded and cold-blooded animals in, for example, the energetic cost of establishing a new individual?

Similar questions could be posed for all the major activities of organisms, including growth and production (see below), but population survival and increase are amongst the most vital areas of interest.

Relatively little work has been carried out specifically to answer these questions but some of the findings are illustrated in Figs 4.1, 4.2 and 4.3.

Ripening

In some cases (e.g. grasses), plant growth almost immediately results in a product. The timing of harvesting is then chiefly related to the quantity available, although improvement in quality may still be possible.

Forage grazed by animals is a good example of this. Regrowth of leaves may provide a satisfactory diet for the grazing animal within a few days of a previous defoliation; it may even be especially attractive at this stage. The total quantity harvested per annum, however, may be much less than if fewer harvests were taken, leaving longer intervals between defoliations. In some extreme circumstances, severe defoliation may not merely reduce plant production, but may kill the plant. Furthermore, frequent defoliation may make inefficient use of plant nutrients, especially nitrogen, since more may be taken up in the early stages of growth than is immediately required. The result is that the amount of plant growth per unit of nitrogen is reduced and the concentration of nitrogen in the plant may be unnecessarily high from the point of view of the requirement of the animal.

There are many such cases, where time of harvesting is primarily related to the achievement of a satisfactory yield/quality relationship (see Figs 4.4 and 4.5).

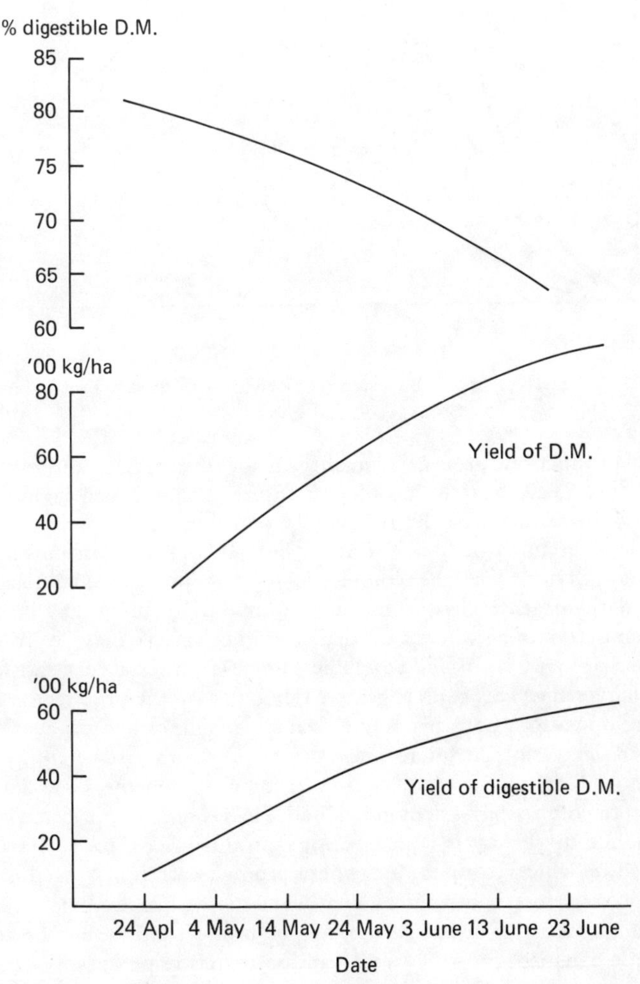

FIGURE 4.4 Perennial ryegrass, seasonal changes in D.M. yield and digestibility (after Aldrich and Dent, 1967)

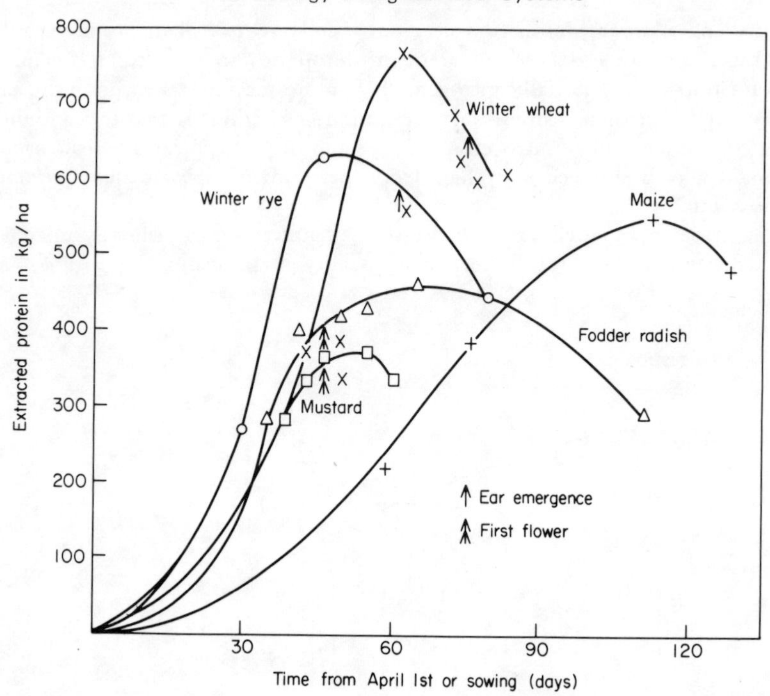

FIGURE 4.5 The effect of age on the protein yield of different crops (from Arkcoll, 1971).

In other production processes, quality is of over-riding importance and harvests are timed very exactly in order to ensure it: this is well exemplified by the pea crop. Sometimes the desired quality depends upon ripening processes. This is so for all grain to be used for direct human consumption and the timing of harvesting has to await the completion of the process. The relationship between climatic/weather conditions and the time of ripening may be of great importance in determining where a particular crop or variety may be grown.

In the case of many fruits and vegetables, however, ripening can satisfactorily occur *after* harvesting. This confers greater flexibility on the producer in relation to harvesting operations and this may be valuable, simply because weather and soil conditions may not be suitable for these operations just when the crop is ripe. Since not all fruit or seeds of a crop become ripe on the same day, and it may not be possible to pick them individually, losses may occur if one waits for the last product to ripen. Of course, unripe products have to have a safe and suitable place in which to ripen, but where products are transported over long distances the transport container may provide such a place at no extra cost. Indeed, as in the case of bananas, the cost of transportation may be reduced, because unripe bananas require less refrigeration than do ripe ones.

The biological processes involved in ripening are therefore of considerable consequence and it is quite possible to increase the efficiency of production per unit of land by timeliness of harvesting or by selecting plant species that ripen quickly or at a particular time.

Individual harvesting of fruits or seeds might also increase the efficiency with which some resources are used but may not be possible, practicable or economic.

"Maintenance"

It is less usual to think about "maintenance" in plants than it is in animals, but there are good reasons for doing so and, indeed, it could even be argued that the concept has more relevance to crops than to animals.

The difficulties of applying the idea to growing or productive animals will be discussed in the next chapter. It is relatively easy to see its application to non-productive animals, however. If an animal has to be fed, at some cost, simply to maintain it alive, healthy and without weight loss, during periods when it is producing nothing, then the size of this cost matters. Thus, the amount of food needed per unit weight of animal is important and this is one way of considering the "maintenance requirement".

The same sort of process occurs in plants, but the costs appear less, or at any rate less obvious.

At low temperatures, for example, when growth may be negligible (or even negative), resources are still being used by the plant. It may be argued that these resources are mainly water, radiant energy, CO_2 and nitrogen and that they do not generally incur a monetary cost. Alternatively, for non-leguminous plants, it may be argued that the soil nitrogen used is returned to the soil following death of foliage and roots (if they are not, in fact, being harvested). This ignores any increase in the rate of gaseous loss due to such recycling. If such an increase occurred and was large enough to be important, it might be better to have a more completely dormant plant species.

More important, however, is the validity of the assumption that the other resources are free: the cost of the land really reflects their availability, although land prices may also be greatly influenced by many other factors as well.

Thus, it may be argued that wherever resources are used and do not result in crop production, they could be used more effectively by other plant species, and clearly this is often possible. The alternatives may also incur their own additional costs (of cultivation, for example), however, and all these have to be taken into account. Alternative usage may not necessarily be envisaged at the same time, since the advantage of not having a non-productive plant cover may lie in resource conservation for use in another period of the year. This may be imagined for nitrogen and water, and the use of fallowing mentioned earlier is an example of this.

It may be worth considering just how much of the resources used by plants are, in fact, used primarily to maintain the plant organism, as opposed to contributing to any increase in its size, and whether there are major differences in this respect between different plant species.

Of course, all living organisms are continually changing and tissues are replaced at varying rates. So it is not really possible to envisage a plant that only uses resources to maintain exactly the same living structure. Nevertheless, in agricultural terms, the concept of "maintenance", as a quantity of resources required when no production occurs, is extremely useful and may be of value in thinking about crop production, as it has been with animals.

In the case of crops, some of the factors that cause the requirement to vary in animals, such as physical activity, are not relevant, but there may nevertheless be considerable differences between species.

Senescence and Death

One of the most important consequences of the continuous change occurring in living organisms is that growth is accompanied by death: new cells are added and others die and are shed. Structurally, and perhaps more so in plants than in animals, it is often important that dead cells are not shed but contribute to the strength of the organism, without increasing its maintenance requirement. This latter point does not necessarily have the same significance in animals, because they move about. Thus, an elephant that incorporates dead cells into massive tusks incurs the energetic cost of carrying them about but a tree with a massive trunk benefits from the environmental volume thus made available to its living tissues without incurring any additional cost.

The tree happens also to be a good example of storage of products after growth and death have occurred. In the case of foliar production, only net growth contributes and substantial losses due to senescence and death may occur between harvests. Table 4.16 shows the estimated cumulative transfers of carbon between the atmosphere and different parts of trees in an oak forest in America.

This continuous senescence is more obvious in plants, at least in the above-ground parts, than in animals. Death of the individual, on the other hand, tends to be more obvious in animals, or at any rate in the larger ones. Perhaps size is, in fact, the major factor: it would be difficult to ignore either a dead elephant or a dead tree, whereas dead insects and dead seedlings or algae make little impression.

Production

At the beginning of this chapter, we considered the range of plant products but little has been said about the productivity of the crop species involved. There is no real difficulty in comparing the productivity of species that produce similar products and use similar resources: thus plant varieties are relatively easy to compare, for specified purposes.

Indeed, it can be argued that if it is tomatoes that are wanted, there is no conceivable point in comparing the tomato plant with, for example, the potato.

Many plants, however, produce raw materials for processing and *can* be compared in terms of the amount of material suitable for processing that they produce per unit of any specified resource.

At its most general, this proposition regards the human consumer as a processor and we may then consider the relative efficiency with which different

TABLE 4.16 Estimated cumulative annual transfers of carbon (g m^{-2} $year^{-1}$)[a] in an oak-pine forest at Brookhaven National Laboratory, Long Island (after Woodwell and Whittaker, 1968; Woodwell and Botkin, 1970). From Olson (1970).

Source	Pool index	Destination 1 Atmosphere	2 Leaves	3 Flowers and fruit	4 Bole and branch	5 Underground parts	6 Substrate	7 Consumers
Atmosphere	1	—	166[b]	~0	627–720[c]	371[c]	~0	0
Leaves	2	?[e]	—	0	50[d, e]	0[e,f]	122	14
Flower + fruit	3	?	0	—	0	0	11	?
Bole + branch	4	27–510	20[c]	11[d]	—	0[c]	32	?
Underground parts	5	190–	?[f]	?[f]	?[f]	—	140	0
Substrate + decomposers	6	280	0	0	0	0	—	?
Consumers (above-group)	7	12	0	0	0	0	0	—

a Assuming 45% for tree and herb leaves, roots and organic substrate; 47% for stems; and 50% for flowers and fruits. Slight discrepancies may result from wrong selection of these factors, and from differences in dry matter estimates between Woodwell and Whittaker (1968) and Woodwell and Botkin (1970).

b Leaf litter 122 + consumptions 14 + loss before abscission 39 + reproductive 11 = 186, but 20 g m^{-2} $year^{-1}$ of this is assumed to consist of translocation from stems: see *c* below.

c Carbon for stem and root growth, while photosynthesized in leaves, is assumed to be promptly translocated to stem and root cambium and to a labile storage pool. The assumed 20 g in stems seems minimal for the spring flush of growth.

d Carbon for reproduction is shown as temporarily translocated into stems and then into flowers and fruit, with a probable value of 11 g shown here. (Actual amounts should vary yearly as a function of prior fruiting and weather.)

e Loss of 39 g carbon by leaves before abscission is here shown going to storage in stems, but might be partly due to leaf respiration (not estimated here) and to leaching loss: 50 = 39 + 11 (from *d*).

f Additional stem and root storage might be expected in order to balance upward spring translocation during normal flushing of new growth, most conspicuously during sprouting after removal of tops. Estimates are not available for these cumulative transfers.

crops produce raw materials for direct human use. Not all of them make their contribution through human food but this is certainly a major route and one likely to be important for a very long time.

The most useful basis for comparison of food crops is either in terms of the efficiency of energy or protein production. The disadvantages of this are that man does not live by either alone and requires a proper balance, not only between them but between the constituent amino acids of the proteins. Table 4.17 illustrates typical amino acid proportions in crop and animal products, compared with the dietary requirements of chicks and pigs.

TABLE 4.17 The requirements for amino acids compared with the proportions present in certain feed proteins (from Pyke, 1971)

Amino acids	Requirement (% of diet)		Composition of protein (g/16 g N)					
	Chicks	Pigs	Wheat	Maize	Sorghum	Egg	Milk	Meat
Histidine	0·3	0·3	2·1	2·5	1·8	2·4	2·6	3·4
Isoleucine	0·6	0·6	4·1	6·4	5·2	5·7	7·5	5·4
Leucine	1·4	1·2	6·8	15·0	13·2	8·8	11·0	5·1
Lysine	1·0	1·1	2·7	2·3	2·6	7·2	8·7	10·1
Methionine	0·4	0·4	2·0	3·1	0·5	3·8	3·2	2·6
Phenylalanine	0·8	0·7	5·0	5·0	4·6	5·7	4·4	4·4
Threonine	0·6	0·6	3·0	3·7	3·1	5·3	4·7	5·1
Tryptophan	0·2	0·2	1·3	0·6	1·0	1·3	1·9	1·1
Valine	0·8	0·6	4·3	5·3	4·7	8·8	7·0	7·0
Arginine	1·2	0·3	4·3	4·8	3·2	6·5	4·2	6·9

The simplest way is to consider energy and protein separately, on the grounds that they are both major dietary needs of man, and that no-one lives on the produce of one crop or even on crop products alone.

There are many ways of assessing the efficiency of crop production (Acock *et al.*, 1971; Monteith, 1966; Gibbon *et al.*, 1970) and the only essential is that any expression must be justified by the purpose of stating it and any calculation justified by the purpose of making it.

In general terms, efficiency (E) can always be described as an output (O) per unit of some input (I), over a specified period of time and in a specified context (see Spedding, 1973).

Thus:

$$E = \frac{O}{I}$$

and, if O is to be the output of energy or protein, the other main decision has to be the terms in which I is to be described. It could be legitimately expressed as energy, protein, water, land, labour, money or any of the other resources used including time.

There are very good reasons, furthermore, for splitting energy into two categories:

(a) that derived from current solar radiation, and
(b) the "support" energy derived most commonly from "fossil" fuels (i.e. the products of past solar radiation).

The most usual expression and probably the most useful single one, however, is per unit of land.

A convenient time period is one year for those crops that produce annually, but longer periods are needed to assess such crops as rubber, cocoa and cloves, where the plant has a long establishment period before it starts producing (see Table 4.18). Some of these perennial tree crops may continue production for many years, however, and the establishment period may be calculated as spread over these years.

TABLE 4.18 The production periods of tropical perennial crops (after Ruthenberg, 1971)

	Years to first crop	Years of production
Field crops		
Sugar cane	1–1·5	4–6
Bananas	1–2	5–50
Pineapples	1·5	3–5
Sisal	3	8
Shrubs		
Coffee	3	12–50
Tea	3	50
Trees		
Oil-palm	3–4	35
Rubber	4–7	35
Cocoa	8–11	80–100
Coconuts	4–6	60
Cloves	8–9	100

The context of the calculation needs very careful consideration: yields differ greatly according to whether or not the crop is being grown in its optimum environment and according to the supply of water and fertilizers.

Indeed, Monteith (1972) has summarized the factors affecting the efficiency with which plants store solar energy as follows:

1. latitude and season;
2. cloudiness and aerosol content of the atmosphere;
3. the spectral composition of radiation;
4. the quantum need of the photochemical process;
5. leaf area index and leaf arrangement;
6. the concentration of CO_2 in the canopy and the diffusion resistance of individual leaves;
7. the fraction of assimilates used in respiration.

TABLE 4.19 Provisional geographic estimates for ecosystem areas, annual net primary carbon production and carbon budget of the world's land masses (from Olson, 1970)

	Area	Net primary production of carbon (NPP)		Live carbon pool		NPP/pool
	$10^6 km^2$	$tons/km^2/year$	10^9 tons	$tons/km^2$	10^9 tons	
Woodland or forest						
temperate "cold-deciduous"	8	1000	8	10 000	80	0·10
conifer boreal and mixed	15	600	9	8000	120	0.075
rainforest: temperate	1	1200	1·2	12 000	12	0·10
rainforest: tropical, subtropical	10	1500	15	20 000	200	0·075
dry woodlands (various)	14	200	2·8	5000	70	0·04
Subtotal	48		36		482	
Non-forest						
agricultural	15	400	6	1000	15	0·4
grassland	26	300	7·8	700	18·2	0·43
tundra-like	12	100	1·2	600	7·2	0·17
other "desert"	32	100	3·2	600	19·2	0·17
glaciers	15	—	0			
Subtotal	100		18·2		59·6	
Continents						
Average/continents	148		54·2		541·6	
Average/earth surface		366		3800		
(continental share)		106		1100		

TABLE 4.20 Total quantities of nutrients in the vegetation of oak* (from Reichle, 1970)

Forest	Soil t/ha	Aerial parts kg/ha	Underground parts kg/ha	Total vegetation kg/ha	Yearly uptake kg/ha
Virelles	K 26·8	245	97	342	69
	Ca 133	868	380	1248	201
Biomass, 156 t/ha	Mg 6·5	81	21	102	19
Productivity, 14·4 t/ha	N 4·5	406	127	533	92
Soil weight, 1360 t/ha	S —	51	30	81	13
	P 0·9	32	12	44	6·9
Wavreille	K 160	493	131	624	99
Biomass, 380 t/ha	Ca 33·3	1338	310	1648	129
Productivity, 14·3 t/ha	Mg 50·1	126	30	156	24
Soil weight, 6318 t/ha	N 13·8	947	313	1260	123
	P 2·2	63	32	95	9·4

* forest ecosystems of Virelles and Wavreille (Belgium)

Before focusing on food crops, it is worth considering the magnitude of world plant production.

Olson (1970) has estimated the net primary production of carbon and the live carbon pool for the major land masses of the world (see Table 4.19). The enormous carbon production of woodlands and forests and the immense carbon pool they represent emerge very clearly from this table.

The distribution of woodland biomass above and below ground, in terms of mineral elements, is illustrated by the data relating to oak in the forest ecosystems of Virelles and Wavreille in Belgium (Table 4.20).

Energy Production

The energy produced per unit area of land by food crops varies between species and from one latitude to another: the range of values is illustrated in Table 4.21. These have been arranged to show whether energy production is

TABLE 4.21 Range of comparative mean yields (100 kg/ha), for 1962/65, in different parts of the world (after Penman, 1968)

Crop	Yield range
Wheat	7·0–44·9
Maize	9·9–40·7
Rice	15·1–60·5
Potatoes	76–307
Sweet potatoes and yams	50–201

generally greater from one part of the plant than another. It may seem obvious that harvesting the entire plant would yield more than harvesting roots, leaves, seed or fruit alone, but there are two major drawbacks. First, at harvest times, many plants have diverted a very high proportion of their total energy into the product part or organ and the energy expended in harvesting the remainder might be disproportionately high. Secondly, harvesting the whole plant would destroy the usefulness of perennials and those plants which can be harvested several times in a season (e.g. forage grasses and tea).

The advantages of perennials ought to be that, since the plant is present the whole time, advantage can be taken of the entire growing season.

Certainly many annuals and biennials appear to waste a considerable part of the solar radiation available in the early part of the year (Watson, 1971), simply because they have not yet developed sufficient leaf.

Nevertheless, the annual energy production of annuals may be quite as high as that of perennials. Furthermore, in some areas, a succession of annual crops of different species can be sown within one year, with the possibility of tailoring particular crops to particular climatic periods of the year.

Another possibility, for both annuals and perennials, is to mix different species in the same population, either completely interspersed or in alternate rows (as in intercropping). This may result in a higher yield than growing each species separately. An example of this was found for mixtures or maize and beans (*Phaseolus vulgaris*) at Makerere (Willey and Osiru, 1972). The mixture yielded up to 38% more than could be achieved by growing the crops separately. This is relevant where both crops are required but does not *necessarily* produce a greater energy yield than the higher-yielding crop alone. Nevertheless, it has been demonstrated that mixtures *can* result in an increased harvestable yield of energy over the best single species crops.

The efficiency with which solar energy is converted to energy in harvested crops varies with plant species, nutrient supply, climate and plant population. Equations have been proposed (Gibbon *et al.*, 1970) which can be used to derive estimates of the ceiling yield of a crop from a measure of the plant population and the level of fertilizer applied.

Such calculations show that optimum plant populations may vary with radiation level in some species and that both potential yield and efficiency vary greatly with site.

Using an energy content of 4 kcal/g of D.M. for all crops, Gibbon *et al.* (1970) calculated energetic efficiency (E) from the following equation:

$$E = \frac{\text{Total energy content of plant dry matter} \times 100}{\text{Total solar energy available}}$$

Values for E for maize (var. Caragua) varied from 0·88 at Headley (England) to 3·12 in Turin. Kale varied from 1·12 (Rome) to 1·79 (Headley) and sugar beet from 1·79 to 1·97 at different sites in England.

It should be noted that the calculation of efficiency may be based on either "total" (i.e. all wavelengths) or "usable" (i.e. wavelengths between 0·4 μm and 0·7 μm) solar radiation and that the value of E varies with the environmental context.

Furthermore, it is essential to define the period over which the calculation is made. Those quoted above were calculated from mean crop growth rates and mean radiation receipts over the same period. It may seem sensible to calculate efficiencies for annuals only when they are present but the efficiency of energy use on an annual basis is not necessarily indicated. For perennials, an annual basis for the calculation would seem more appropriate, even though little or no growth would be taking place over substantial periods.

For the same site, this is virtually the same as a straight comparison of yields of energy per hectare. Holmes (1971) gives an example of this for different crops grown under U.K. conditions (Table 4.22).

Penman (1971) has estimated the theoretical upper limit of efficiency of energy fixation by cereals as between 8 and 10% of total solar radiation during the life of the crop, compared with values as low as 0.04-0.1% achieved in subsistence farming.

Monteith (1966) has calculated a theoretical maximum conversion of available visible radiation of 18%, but this is only obtainable at low light inten-

TABLE 4.22 Energy yields for crops in U.K. (after Holmes, 1971)

Crop	Yield of Energy (Mcal/ha)
Wheat	14 300
Barley	12 000
Vining peas	3100
Brussels sprouts	4000
Cabbage	8100
Potatoes	24 000
Sugar beet	27 000

sities. At normal light intensities, the maximum is in the 5-6% range (Roberts, 1974). Compared with this, and for the periods of active growth alone, Thorne (1971) calculated values of:

2·0% for cereal grain (mid-June–mid-August)
1·5% for beet sugar (June–October)
2·4% for potato tubers (June–September).

There appears to be considerable scope for further improvement in the fundamental process of making agricultural use of solar radiation.

Protein Production

Proteins are distributed throughout the plant, but may be concentrated in particular organs, such as seeds, making harvesting easier. In leaves in which protein may be present in low concentration, largely as enzymes, it is now possible to extract 55-75% of the total crop nitrogen from most species, unless they are very old and woody (Arkcoll, 1971). There are, therefore, several different ways of measuring protein production from crops and the possible methods of processing include consumption by animals of various kinds. These possibilities are shown diagrammatically in Fig. 4.6.

Typical yields of seed protein are given in Table 4.23 and yields of total protein in the leaves of grasses and legumes for animal feeding are shown in Table 4.24. Protein production in roots and tubers is illustrated in Table 4.25.

Leaf protein extraction (see Pirie, 1971) may relate to leafy crops grown specially for the purpose or to foliar by-products of root, grain, seed or tuber crops. Production from the former is, of course much greater (see Table 4.26), but the amounts harvested from unwanted by-products are substantial (see Table 4.27).

Indeed, the extracted protein yield of 250 kg/ha from potato haulm is greater than the annual *ca.* 60 kg that might be expected per hectare from beef production and approaches the protein yield of dairy cows (*ca.* 300 kg/ha).

One of the interesting features of the work on leaf protein extraction is that research is being carried out on plant species not hitherto utilized in agriculture (Hollo and Koch, 1971). Some species of such genera as *Amaranthus, Atriplex*

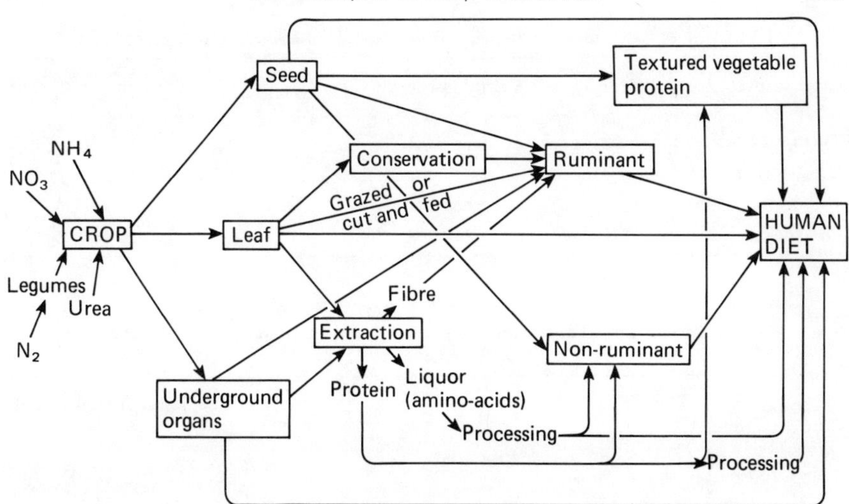

FIGURE 4.6 Diagrammatic representation of protein production from crops.

TABLE 4.23 Protein production from seed crops in the U.K. (after Scott, 1974)

Crop	Yield of crude protein kg/ha
Wheat	425
Barley	327
Oats	368
Winter beans	531
Spring beans	527
Winter rape	478
Spring rape	355
Peas	461

TABLE 4.24 Yields of protein from the main pasture grasses and legumes in the U.K. (data from Spedding and Diekmahns, 1972)

Species		Digestible crude protein[a] kg/ha
Grasses	*Lolium perenne*	1160
	Lolium multiflorum	1600
	Phleum pratense	1330
	Dactylis glomerata	1360
	Festuca pratensis	1190
	Festuca arundinacea	1350
Legumes	*Trifolium repens*	1960
	Trifolium pratense	1580
	Medicago sativa	2000
	Onobrychis viciifolia	1710

[a] D.C.P. = N x 6·25

TABLE 4.25 Protein production by roots and tubers (from Rodgers, 1974)

Crop	% Crude protein (C.P.)	Yield of C.P. (kg/ha)
Turnips (stockfeed)	1·2	610
Sugar beet (root)	1·1	360
(tops)	2·0	650
Potatoes	1·9	480

TABLE 4.26 Total annual yields of leaf protein achieved at Rothamsted (after Arkoll, 1971)

Crop sequence	Total annual protein yield (kg/ha)
Wheat, fodder radish	1244
Wheat, mustard	1261
Wheat + regrowth, fodder radish (1967)	1448
Wheat + regrowth, fodder radish (1968)	2014
Rye, mustard	1373
Red clover + regrowths (irrigated)	1247
Cocksfoot + regrowths	1670
Lucerne + regrowths	1106

TABLE 4.27 Protein yields from by-product leaves at Rothamsted (Sources: Pirie, 1969; Arkcoll, 1967; Arkcoll and Festenstein, 1969)

Crop	Extracted protein (kg/ha)
Potato haulm	250
Pea haulm	350
Sugar beet tops	500–830
Maize, after removal of cobs at milky stage	300

TABLE 4.28 Protein content and extractability in various plant species (Source: Joshi, 1971)

		Protein extracted	
Plant species	N in D.M. %	N as % of total crop N	kg/ha
Lablab niger	4·0	48	375
Dolichos uniflorus	3·63	49	416
Amaranthus paniculatus	3·71	59	583
Beta vulgaris	3·0	52·5	87
Sesbania sesban var. *picta*	4·0	56	342
Daucus carota var. *sativa*	4·1	36	349
Raphanus sativus	3·96	54	271

and *Chenopodium* show favourable qualities. In some cases the content of lysine and methionine is high; this is important since methionine is the limiting amino acid in all leaf protein preparations. Other important attributes are protein content and the extractibility of the protein; both vary between plant species (see Table 4.28).

Another kind of by-product occurs where aquatic weeds have to be removed to keep waterways clear. Several of these species have been investigated. Emergent plants seem much more promising than either submerged or floating-leaved plants, from the point of view of protein yield, although floating plants such as water hyacinth (*Eichhornia crassipes*), *Lemna* and *Azolla* offer advantages in terms of harvesting. Maximum protein yields of 590, 478 and 362 kg/ha were obtained with *Justicia americana, Alternatherea philoxeroides* and *Sagittaria latifolia*, respectively (Boyd, 1968).

Processing of this kind naturally involves additional inputs of energy, in the form of fuel, and it is considered that the electricity requirement for mechanical disintegration is about twice that of conventional drying procedures. However, the specific calorie consumption (of 300-350 kcal to vaporize 1 kg of water) is thought to be much less than that involved in hot air drying equipment (900-1000 kcal/kg). The production figures from a pilot plant (Hollo and Koch, 1971), showed that 45-55 kWh of electricity and about 40 kg of fuel oil were required to process one ton of green matter.

The Use of Resources

Productivity has been discussed here chiefly in terms of energy or protein output per unit area of land (or water). It has to be remembered that there are many other important resources and the efficiency with which they are used may be more important even than the efficiency of land use. Indeed, from an economic point of view, this may often be so.

In animal production (see Chapter 5), the cost of feed is a very high proportion of the total costs and, since the feed is mostly produced from the land, land use is of paramount importance.

In crop production, however, it is often labour or machinery that contributes the major part of the costs. This is particularly true, of course, for crops such as tea, where about 4000 shoots have to be picked for 1 kg of made tea. Ruthenberg (1971) estimates that *ca.* 80% of the costs of establishing a plantation without a factory and *ca.* 60% of the costs of a mature plantation are wages and salaries. In Ceylon, 44% of the labour is calculated to be spent on harvesting the leaf, 19% on hoeing weeds and 8% on work in the factory.

In some crops the provision of seed is a major item and there are substantial differences in the production of yield that has to be retained (or its equivalent in seed etc.) in order to establish another crop (see Table 4.29).

The calculation of efficiency has to be related to the importance of the resource employed and improvements in the efficiency with which one resource is used have to take into account the consequent effects on other aspects of efficiency.

TABLE 4.29 Proportion of yield needed to establish another crop (based on calculations by R. V. Large, Grassland Research Institute)

	Yield (Y) (kg/ha)	Seed required (S) (kg/ha)	$\frac{S}{Y} \times 100$
Potatoes	25 000	2500	10
Spring barley	3600	170	4·7
Winter wheat	3700	120	3·2
Spring oats	3900	170	4·4
Winter beans	2000	230	11·5
Winter peas	3100	170	5·5
Maize	5000	10	0·2

The same argument applies to discussions of crop production potential. Ideally, potential production should be related to each of the most important resources.

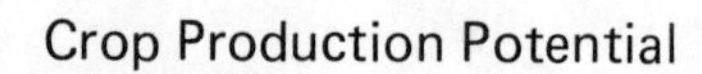

Crop Production Potential

As Williams (1972) has pointed out, approximately 70% of the food needs of mankind are provided directly by plants in the form of grain, tubers, roots, sugar, vegetables and fruit. This dependence on crops, rather than animals, is even greater in the underdeveloped world, where the majority of the world's population live and where crop yields tend to be lower.

Unfortunately, translation of the methods currently used to sustain very high crop yields in developed countries, directly to the less-developed areas, may not be practicable. Furthermore, it may not even be desirable and it certainly implies the use, on an enormous scale, of inputs of fertilizer, weed-killers, machinery and capital. One effect is often to reduce the need for labour in areas where a major function of agriculture is to provide employment.

However, it is often pointed out that current crop production only manages to accumulate in the form of products less than 1% of the available light energy, and substantial improvements may be expected (Roberts, 1974).

It is essential that such advances be of a form that can benefit the community as a whole and do not involve procedures resulting in problems due either to the inputs required or the outputs produced.

References

Aldrich, D. T. and Dent, J. W. (1967). *J. natn. Inst. agric. Bot.* **11**, 104-13.

Acock, B., Thornley, J. H. M. and Wilson, J. W. (1971). *In* "Potential Crop Production" (P. F. Wareing and J. P. Cooper, eds), p. 43. Heinemann Educational Books, London.

Arkcoll, D. B. (1967). *Rep. Rothamsted exp. Stn,* 1966, p. 104.

Arkcoll, D. B. (1971). *In* "Leaf Protein". IBP Handbook No. 20 (N. W. Pirie, ed.), Chapter 1. Blackwells, Oxford and Edinburgh.

Arkcoll, D. B. and Festenstein, G. N. (1969). *A. Rep. Rothamsted Exp. Stn* 1968, p. 119.

Arnott, R. A., Brockington, N. R. and Spedding, C. R. W. (1974) *J. exp. Bot.* (in press).

Boyd, C. E. (1968). *Econ. Bot.* **22**, 359.

Brown, L. R. and Finsterbusch, G. W. (1972). "Man and his Environment: Food". Harper and Row.

Cooke, G. W. (1967). "The Control of Soil Fertility". Crosby Lockwood and Son Ltd., London.

Cooper, J. P. (1970). *Herb. Abstr.* **40** (1), 1.

de Wit, C. T. (1967). *In* "Harvesting the Sun" (A. S. Pietro, F. A. Greer, and T. J. Army, eds), pp. 315-320. Academic Press, New York and London.

Dittmer, H. J. (1972). "Modern Plant Biology". Van Nostrand Reinhold, New York.

Donald, C. M. (1960). *J. Aust. Inst. agric. Sci.* **26**, 319-338.

Floate, M. J. S. (1970). *J. Br. Grassl. Soc.* **25** (4), 295-302.

Gadgil, M. and Solbrig, O. T. (1972). *Am. Nat.* **106** (947), 14-31.

Gibbon, D., Holliday, R., Mattei, F. and Luppi, G. (1970). *Exp. Agric.* **6**, 197-204.

Harper, J. L. (1967). *J. Ecol.* **55**, 247-270.

Harper, J. L., Lovell, P. H. and Moore, K. G. (1970). *A. Rev. Ecol. Syst.* **1**, 327-356.

Hoare, E. R. (1967). *Outlook on Agriculture* **5** (4), 139-143.

Hollo, J. and Koch, L. (1971). *In* "Leaf Protein". IBP Handbook No. 20 (N. W. Pirie, ed.), Chapter 6. Blackwells, Oxford and Edinburgh.

Holmes, W. (1971). *In* "Potential Crop Production" (P. F. Wareing and J. P. Cooper, eds), p. 213. Heinemann Educational Books, London.

Hutchinson, J. B. (ed.) (1969). "Population and Food Supply", Chapter 8. Cambridge University Press.

Janzen, D. H. (1969) *Evolution* **23**, 1-27.

Joshi, R. N. (1971). *In* "Leaf Protein" IBP Handbook No. 20. (N. W. Pirie, ed.), Chapter 2. Blackwells, Oxford and Edinburgh.

Kipps, M. S. (1970). "Production of Field Crops" (6th edition) McGraw-Hill.

Lieth, H. (1972). *Angew. Bot.* no. 1 (Symp. German Bot. Soc. in Innsbruck, 1971).

MacArthur, R. H. (1972). "Geographical Ecology". Harper and Row, New York.

McKee, H. S. (1962). "Nitrogen Metabolism in Plants". Oxford University Press.

Monteith, J. L. (1966). *Exp. Agric.*, **2**, 1-14.

Monteith, J. L. (1972). *J. appl. Ecol.* **9**, 747-766.

Nutman, P. S. (1972). *Rep. Rothamsted exp. Stn,* 1971, Pt. 1, 91-100.

Ogden, J. (1968). Ph.D. thesis, University of Wales.

Olson, J. S. (1970). *In* "Analysis of Temperate Forest Ecosystems" (D. E. Reichle, ed.) Ecological Studies, Vol. 1. Springer, Berlin-Heidelberg-New York.

Penman, H. L. (1968). Proc. Cent. Symp. Faculty of Agric. Sci. Amer. Univ., Beirut, Lebanon. pp. 333-353.

Penman, H. L. (1971). *In* "Potential Crop Production" (P. F. Wareing and J. P. Cooper, eds), p. 89. Heinemann Education Books, London.

Pirie, N. W. (1969). *Proc. II Int. Bot. Congr.*, Seattle, Wash.

Pirie, N. W. (1971). *In* "Leaf Protein", IBP Handbook No. 20 (N. W. Pirie, ed.), Chapter 3. Blackwells, Oxford and Edinburgh.

Pyke, M. (1971). *In* "Potential Crop Production". (P. F. Wareing and J. P. Cooper, eds), Chapter 13. Heinemann Educational Books, London.

Reichle, D. E. (ed.) (1970) Analysis of Temperate Forest Ecosystems. Ecological Studies, Vol. 1. Springer, Berlin-Heidelberg-New York.

Roberts, E. H. (1974). *In* "Human Food Chains and Nutrient Cycles" (A. N. Duckham and J. G. W. Jones, eds), Chapter 6. Elsevier, Amsterdam.

Rodgers, B. M. (1974). Personal communication.

Russell, E. W. (1961). "Soil Conditions and Plant Growth" (9th edition). Longmans, London.

Ruthenberg, H. (1971). "Farming Systems in the Tropics". Oxford University Press.

Salisbury, E. J. (1942). "The Reproductive Capacity of Plants". Studies in Quantitative Biology. G. Bell and Sons, London.

Schery, R. W. (1972). "Plants for Man". (2nd edition) Prentice-Hall, New Jersey.

Scott, R. K. (1972). *Agric. Progr.* **47**, 139-153.

Scott, R. K. (1974). *Proc. Reading Univ. Agric. Club Conf.* 1974. 75-88.

Scott Russell, R. (1971). *In* "Potential Crop Production" (P. F. Wareing and J. P. Cooper, eds). Heinemann Educational Books, London.

Spedding, C. R. W. (1973). *In* "The Biological Efficiency of Protein Production" (J. G. W. Jones, ed.), Cambridge University Press.

Spedding, C. R. W. and Diekmahns, E. C. (1972). "Grasses and Legumes in British Agriculture", Commonwealth Agricultural Bureaux, Farnham Royal, Bucks.

Stevens, O. A. (1932). *Am. J. Bot.* **19**, 784-794.

Stevens, O. A. (1957). *Weeds* **5**, 46-55.

Thorne, G. N. (1971). *In* "Potential Crop Production" (P. F. Wareing and J. P. Cooper, eds), p. 143. Heinemann Educational Books, London.

Watson, D. J. (1947). *Ann. Bot.* **11**, 41-76.

Watson, S. J. (1971). *Advmt Sci., Lond.* **27**, 1970-71. pp. 1-13.

Willey, R. W. and Osiru, D. S. O. (1972). *J. agric. Sci. Camb.* **79**, 517-529.

Williams, Watkin (1972). *In* "Growing points in Science", p. 142, HMSO, London.

Winter, E. J. (1967). *Outlook on Agriculture* **5** (4), 144-148.

Woodwell, G. M. and Whittaker, R. H. (1968). *In* Symposium on Primary Productivity and Mineral Cycling in Natural Ecosystems (H. E. Young, ed.), pp. 151-166. University of Maine Press, Orono.

Woodwell, G. M. and Botkin, D. B. (1970). *In* "Analysis of Temperate Forest Ecosystems" (D. E. Reichle, ed.), Chapter 7, pp. 73-85, Vol. 1. Ecological Studies, Springer, Berlin-Heidelberg-New York.

5 The Principles of Animal Production

Animal production is undertaken for purposes which must be consistent with those of the agricultural system of which it is a part, but which may nevertheless not be the same. This is simply because animal production, in the biological sense, has a role *within* agricultural systems, just as is the case with crop production.

It is useful to consider first, then, the particular purposes of animal production.

Purposes of Animal Production

These may be grouped under six main headings:

1. Output—production and performance;
2. Collection of plant food;
3. Conversion of nutrients;
4. Concentration of nutrients;
5. Elimination of toxic materials;
6. Continuity of food supply.

1. Output

The variety of products, by-products and contributions in the form of performance that are represented by the outputs of animal production, were listed in Chapter 1 (see Table 1.2).

The animal species and breeds currently used in world agriculture in substantial numbers are indicated in Table 5.1. Tables 5.2 to 5.6 show the animals used in the production of milk, meat, other food products, leather and fibres, with some indication of their yields. In addition, there are animals that produce fur and that are farmed to a greater or lesser extent (Table 5.7).

In some cases, only a limited number of different animal species can actually produce a given product or render a particular service. In most cases, however, there are alternatives and, in some cases, a whole range from which to choose.

Thus, if milk is required, the demand may be quite specific as to the animal producing it, or even as to the chemical analysis of the product, but otherwise

TABLE 5.1 The main agricultural animals of the world

Animal	World population[a] 1970/71 (thousands)	Number of breeds[b] (approx.)
Ass	41 914	12
Buffalo	125 412	7
Cattle	1 141 215	247
Goat	383 025	62
Horse	66 312	124
Pig	667 689	54
Sheep	1 074 677	230
Mule	14 733	

Sources:
[a] FAO (1972)
[b] Mason (1951)

TABLE 5.2 Animal species used for milk production in modern agriculture

Animal species	Yield per head (l)		Period of production	Reference
	Typical	Maximum		
Buffalo	900–1364	2728–3182	Year	Turner (1971) Cockrill (1969)
Camel	1350	3600	Year	Leupold (1968)
Cattle	3864	11 365	Year	Baker (1973) Turner (1971)
Ass				Turner (1971)
Horse	11–17		Day	I.D.F. Turner (1971)
Goat	600	2045	Year	Mackenzie (1972) French (1970) Turner (1971)
Reindeer				Turner (1971)
Sheep	150–200	1200	Year	Turner (1971) Treacher (1969) Boyazoglu (1963)
Yak				Turner (1971)
Eland	350	630		Crawford (1968) Krutyporoh and Treus (1969)

the animal may be chosen on quite other grounds. Reasons for choice include the ability of the species to live and produce on the diets available, the degree of adaptation to a particular climate or soil type, resistance to disease, and the availability of breeding or young stock.

In the case of most animal products used for human food, there has been little competition with the quite different products produced by crops. This may

TABLE 5.3 Animals used for meat production

Animal species	Typical yield per head kg	Period of production	References
Buffalo	144	14 months	Charles and Johnson (1972)
	279	48 months	Turner (1971)
	250	> 3 years	Young and Van den Hever (1968)
	216	18 months	El Naggar *et al.* (1970)
Alpaca			
Camel			
Llama			
Cattle	250	16 months	Turner (1971)
			Barker (1973)
Dog			
Ass			
Horse			
Goat			
Pig	50	5 months	Turner (1971)
			Lawrie (1970)
Reindeer			
Rabbit	1	9 weeks	Turner (1971)
			Walsingham (1973)
Guinea-pig			
Sheep	18	16 weeks	Turner (1971)
			Lawrie (1970)
Yak			
Poultry	1·45	9 weeks	Morris (1971)
Deer	82	3 years	Hope (1972)
	113	27 months	Bannerman and Blaxter (1969)
Musk Oxen			

TABLE 5.4 Other Animals used for food production

Products	Animal species	Yield per head		Period of production	References
		Typical	Maximum		
Honey	Bees	11 kg/hive	27 kg+/hive	1 year	Hawthorne (1972)
Fish	Trout	45–180g		9–24 months	U.S. Dept. of Int. (1963)
	Cod	5 kg+		6 years	Waterman (1968)
Eggs	Hens	240	300+	1 year	Morris (1971)
	Ducks	250	344	1 year	Ontario Dept. of Agric.
					McArdle (1966)

TABLE 5.5 Animals used for leather production (Raistrick, 1973)

Animal species	Yield per head (ft^2)
Buffalo	35–50
Camel	30
Cattle	35–50
Ass	15
Horse	35–50
Goat	4·5
Pig	10–12
Reindeer	20
Sheep	6
Yak	30
Elephant	100
Hippo	60
Eland	40
Wildebeest	24
Kudu	20
Impala	8
Kangaroo	10
Ostrich	12

TABLE 5.6 Animals used for the production of fibre

Product	Animal species	Annual yield/ head (kg)	Reference
Wool, textile and other fibres	Sheep	1–9	Spedding (1970)
	Alpaca	1·5–5	French (1966)
	Llama	1·5–3·5	French (1966)
	Vicuna	0·18–0·32	French (1966)
	Huanaco	1–2	French (1966)
	Musk Oxen	2·3–2·7	Dodd (1972)
	Cattle		
	Horse		
	Goat (Cashmere)	0·23	Commonwealth Secretariat (1970)
	Reindeer		
	Rabbit (Angora)	0·23	Commonwealth Secretariat (1970)
	Guinea-pig		
	Yak		
	Camel	2·3–4·5	Commonwealth Secretariat (1970)

TABLE 5.7 Fur-bearing domestic animals

Chinchilla	Hamster
Ermine	Sheep
Fox	Kangaroo
Mink	Calf
Sable	Pony
Rabbit	

change radically with the advent of a wide range of "analogue" foods, based, for example, on textured vegetable protein (Rolfe and Spicer, 1973), although butter and margarine have existed side by side for a very long time (the latter introduced at about the time of the Franco-Prussian war of 1870-71).

With products used for clothing, however, the situation has been rather different. Approximately 54% of the world total fibre production is of plant origin, about 21% comes from animals and about 25% is artificially produced. The relative contributions from different animals and crops are given in Table 5.8.

It is necessary sometimes to compare the relative efficiency with which different animals perform (see later in this chapter) but it will be clear that efficiency may have to be related to all these other facets of suitability to a particular environment.

TABLE 5.8 Products used for clothing

	World production	Year	References
Of animal origin:			
Leather	10 000 x 10^6 ft^2	1965–66	Raistrick (1973)
	4·3 x 10^6 t (wet salted wt)		FAO (1970a)
Fur	–	–	
Wool (clean)	1·6 x 10^6 t	1971	FAO (1971a)
Silk	40 000 t	1970	FAO (1971a)
Mohair (greasy)	26 000 t	1968	Commonwealth Secretariat (1970)
Auchenidae	4 535 t	Estimated annual production	Commonwealth Secretariat (1970)
Total	6 x 10^6 t		
Of plant origin:			
Cotton	12 x 10^6 t	1971	FAO (1971a)
Flax	648 000 t	1971	FAO (1971a)
Rubber	3 x 10^6 t	1971	FAO (1971a)
Total	15·6 x 10^6 t		
From industry:			
Artificial fibres	7·3 x 10^6 t	1968–69	Commonwealth Secretariat (1970)

2. Collection of Plant Food

This is one of the reasons why particular animals may be kept in a given environment, because they are able to collect plant food that would otherwise be either unavailable to man or grossly uneconomic to harvest.

Large areas of low-producing grasslands are a common example, especially if they lie on steep slopes or on land which is rocky, boggy or very uneven. Physical features may make harvesting operations very difficult, and therefore expensive, and low productivity simply means that a great deal of land has to be covered in order to harvest an appreciable amount of crop.

Outside agriculture, the use of marine animals to harvest dilute suspensions of small organisms is a good example: bee-keeping for the production of honey is rather more agricultural and is often engaged in by fruit farmers in order to ensure pollination.

Within modern agriculture, as practised in the more developed countries, animals tend to be less used for this purpose than was at one time the case. The herding of pigs in forests, close shepherding of sheep with men and dogs, the keeping of pigeons to feed off local fields and the keeping of hens as barnyard scavengers, all these have decreased, chiefly because of the high labour demand of such activities, which offer no scope for automation, mechanization or even large-scale operation.

The original idea was quite sound, that food lay around unused and would be wasted unless an animal could be employed to collect it for itself. Barnyard hens found their food very locally and presumably also had a useful role in keeping down the fly population; at the other extreme, pigeons could range over stubble-fields for dropped grain and were not necessarily confined to the fields owned by the pigeon-owner.

Today, the idea of "scavenging", still embodied in stubble-grazing by sheep, is being absorbed into the more controlled concept of the use of by-products. Where profit margins are small, the economics of an enterprise can be significantly influenced by how well by-products are used, and the range and volume of by-products increases as methods of harvesting tend to leave more of the whole crop in the field or, at least, on the farm (see Chapter 8).

Although *collection* of food is a distinguishable activity and can be carried out either by man or animals, no agricultural animals currently collect food except for their own purposes, although they may, of course, transport food that they have not harvested. Bee-keeping for honey production is a marginal example of food collection, often combined with an almost unique service to agriculture, that of pollination.

So the process of collection merges with the next activity to be considered, the conversion of plant material. All these activities have an energetic cost, of course (see Lawton, 1973).

3. Conversion of Nutrients

Wherever plant material can be directly consumed by man, it is generally more efficient to do so, because every animal production process involves substantial losses, especially of energy. This would be true even if we finally consumed the whole animal but, in fact, only a proportion is usually regarded as suitable for human food.

The relative efficiency of using land to produce plants for direct human consumption or to grow plants that are subsequently converted by animals is difficult to determine, since there are so many variables. The fact that the former is generally much more efficient is not in doubt, however, where the two processes can be regarded as genuine alternatives. The situation is summarized in Table 5.9 for a few selected crops and animals, to illustrate the main categories.

Where the alternatives exist, a choice can be made on grounds of profit or product required, but where foods for direct human consumption cannot be produced, one important role of the animal is to convert what *can* be grown into a form that can be used as human food. As will be evident from the preceding section, however, material that can be collected only by using animals falls into the same category, even if it could have been directly used.

Conversion requires digestion processes capable of dealing with these fibrous plant materials and not all animals possess them. Agricultural animals are often grouped, therefore, into those with simple stomachs and those with major adaptations to the alimentary tract, mainly concerned with the digestion of fibre.

TABLE 5.9 Relative efficiency of production per unit of land by crops and animals (from Spedding and Hoxey, 1974)
(The data are shown to indicate orders of magnitude: methods of calculation vary and the sources given should be consulted before making detailed comparisons)

Product Harvested	Protein ($kg\ ha^{-1}\ year^{-1}$)	Energy ($MJ\ ha^{-1}\ year^{-1}$)	Source
Dried grass	700–2200	92 000–218 000	1
Leaf protein	2000	–	2
Cabbage (edible)	1100	33 500	3
Maize grain (North America)	430	83 700	4, 5
Potato (edible tuber)	420	100 400	3
Barley grain (U.K.)	370	62 800	4, 5
Wheat (edible grain)	350	58 600	3
Rice grain (Europe)	320	87 900	4, 5
Cassava roots (Malawi)	[a]246	[a]133 800	4, 6
Rabbit carcass	180	7400	7
Chicken (edible broiler)	92	4600	3
Lamb (edible meat, New Zealand)	62	7500	8
Beef (edible meat, mainly barley-fed)	57	4600	9
Pig (edible meat)	50	7900	3
Lamb (edible meat, U.K. and Eire)	23–43	2100–5400	3, 10
Beef (edible meat, mainly grass-fed)	27	3100	3

[a] kg per hectare per crop (not year)

1. Castle and Holmes (1960)
2. Pirie (1971)
3. Holmes (1970)
4. FAO (1971a)
5. FAO (1970b)
6. Pynaert (1951)
7. Walsingham (1972)
8. Campbell (1968)
9. Duckham and Lloyd (1966)
10. Conway (1968)

The simple-stomached animals include pigs and poultry and, in general, they live on the same kinds of food as could be used by man: they are often said to be in competition with us. However, just because they eat what we, theoretically, could also digest, it by no means follows that we would wish to do so. For example, fish meal may be made from parts of fish that we would not wish to eat; insects and worms may be consumed by birds (e.g. by some game birds) and many other constituents of an animal's diet may be considered unattractive, to say the least. Even when we clearly could consume the food used (e.g. barley), it may be that it could only form a small part of our diet, either because it would be incomplete or because it would become monotonous.

It is nevertheless true that where poor populations have to live off little land, they eat vast quantities of grains, such as rice, and would not use a high proportion of their land to grow food for animals, especially if those animals simply consumed what the human population could eat. Unless they could be extremely intensive, or use additional uncroppable land, however, they would not keep ruminants either. They might be more likely, indeed, to keep a *small* number of simple-stomached animals, such as pigs, to live on household waste.

Animals that are adapted to feed on fibrous diets do not have the necessary cellulose-digesting enzymes themselves, but derive them from symbiotic micro-organisms. The latter are accommodated in sac-like expansions of some

TABLE 5.10 Features of digestive tract in animals

	Length of tract (m)	Empty tract as % of body weight	Weight of stomach + intestines as % of empty body weight	Weight of caecum as % of body weight
Cattle ♀	33–63		5·4	
Sheep ♀	22–43		6·2	0·4
Pig	20–27			
Horse	25–39	6·3	7·0	6·2
Rabbit				2·6
Buffalo ♂		3·8		
Eland ♂		3·5		
Zebra ♂		3·3		0·6
Impala ♂		3·7		
African elephant			13·9	3·4–6·1
Green turtle			7·7	
Goat	22–43		5·3	
Carp			0·9	
Trout			6·9	
Hen			7·2	
Duck (pintail)			14·8	
Camel				0·3

Tables 5.10 and 5.12 based on data from McBee (1971), Nickel *et al.* (1973), Crile and Quiring (1940), Robb *et al.* (1972); Ledger *et al.* (1967).

part of the gut, the most important being the rumen of ruminants (see Chapter 3), the colon of the horse and the caeca of rabbits and birds (such as geese). These serve as reservoirs in which large quantities of food can be stored and digested simultaneously.

Differences in important features of the digestive tract in different species of animals are shown in Table 5.10. There are large differences in the proportion of total body weight that is represented by the alimentary tract, so clearly some species deploy more of their body tissues to digestive activities (Table 5.11). Within the alimentary tract, there are differences in the degree of development of organs such as the stomach, colon and caecum (see Table 5.12). This is illustrated by the volumes of the total tract for the horse and for cattle (200 l and 310 l, respectively) and the volumes of the stomach (10 l and 200 l), the total intestines (190 l and 110 l) and the caecum (30 l and 10 l). There are also differences in total length of the digestive canal. For example, the total length is 30 m in the horse but 50 m in cattle: as a ratio to body length this is 20 : 1 and 30 : 1, respectively (Olsson, 1969).

TABLE 5.11 The alimentary tract as a proportion of body weight

Species	Sex	Empty digestive tract as % of live weight	Reference
Buffalo (*Syncerus caffer*)	♂	3·8	1
Eland (*Taurotragus oryx*)	♂	3·5	1
Zebu bulls (*Bos indicus*)	♂	3·3	1
Zebu fat cows	♀	3·7	1
Impala (*Aepyceros melampus*)	♂	3·7	1
Impala	♀	4·6	1
Horse		6·27	2

Sources: 1. Ledger *et al.* (1967)
2. Robb *et al.* (1972)

TABLE 5.12 Relative capacities (litres) of the alimentary tracts of adult animals

	Rumino-reticulum	True stomach	Caecum	Colon
Pig	–	3·8–8	1·5–2·2	9·0
Cattle	92–197	10–20	10	28
Small ruminants (e.g. sheep)	14–25	1·75–3·3	1·0	5·0
Horse	–	8–18	16–68 Av. 33	55–130 Av. 80

Many of these animals also have a mechanism for masticating the food again later. Ruminants chew the cud and rabbits exhibit refection (the consumption of faeces), so that food may be subjected to digestive processes more than once.

All these features combine to confer some advantages in terms of a reduction in the time needed to collect food and thus a shorter period of exposure to predators. These advantages are particularly valuable for animals grazing large areas, since their function of collecting widely-dispersed material implies substantial periods of grazing.

As a result of structural and functional differences in the alimentary tract, some species digest food more efficiently than others (see Tables 5.13 and 5.14). The digestive efficiencies given have no absolute significance but are taken from an experiment in which different species were compared. The energy available from 1 g of digestible nutrient also varies with the species considered (see Table 5.15).

TABLE 5.13 Mean digestive coefficients (%) of two constituents of a range of concentrate feeds for different species (After Vanschoebroek and Cloet, 1968)

	Crude protein	Crude fibre
Rabbit	61·5	27·5
Pig	71·6	47·2
Sheep	78·1	46·7
Cattle	78·0	40·2

TABLE 5.14 Digestive efficiency of camels, ponies, sheep and elephants (After Hintz *et al.*, 1973)

	% Digestive efficiency (of alfalfa)		
	D.M.	Acid detergent fibre	Crude protein
Camels (1 llama and 1 guanaco)	71·5	61·0	74·7
Ponies	64·8	46·7	65·7
Sheep	63·8	50·5	69·7
Elephant (on natural diet)	43		

TABLE 5.15 Energy content (kcal) per g of digestible nutrient (C.P. = crude protein; C.F. = crude fibre) for different species (After Vanschoebroek and Cloet, 1968)

	Digestible energy		Metabolizable energy		Net energy	
	C.P.	C.F.	C.P.	C.F.	C.P.	C.F.
Rabbit	5·25	4·12	4·30	4·45	2·31	3·16
Pig	5·39	4·15	4·50	4·00	2·40	0·01
Sheep	5·39	3·60	4·08	3·09	1·85	−0·09
Cattle	5·32	3·53	3·63	3·06	1·78	2·37

Ruminants have a further advantage in that they can make use of non-protein nitrogen in their diets. However, it has been concluded (FAO, 1971b) that, while supplements of N-P-N can reduce weight losses of animals fed poor herbages, they can never provide substantial growth or milk production and they have no advantage in supplementing better quality herbages, because the latter already contain sufficient nitrogen.

Animals that can use fibrous foods are the major agricultural converters of foliage to edible meat, milk and eggs. There is no reason why other animals should not be similarly employed, however.

Orraca-Tetteh (1963) has suggested that the giant African snail (*Achatina* spp.) should be cultivated for food, since it can live on a variety of vegetation, even in a decaying state. Large insects, such as locusts, could also be used; indeed, many insects are already eaten but rarely cultivated.

The possibility of finding animals to live on waste products, such as newspaper, is attractive but there are often problems of the accumulation of toxic materials and it is often better to try and incorporate the material into the diet of larger herbivores.

The conversion process usually involves wastage because the animals themselves have to be supported whilst it is going on. The wastage is sometimes deplored, as if it would be better if it could be avoided. At the same time, part of the function of the animal during this conversion process is to extract what is useful to man, to discard what is either not useful or actually harmful, and to concentrate nutrients.

4. Concentration of Nutrients

Herbivores generally have to process very large amounts of plant material in order to extract what they require. We are not equipped to do this, quite apart

TABLE 5.16 Calorific content of various foods (kcal/g) (Source: Taylor, 1969)

	kcal/g
Cabbage (boiled)	0·11
Tomatoes (raw)	0·14
Oranges (with peel)	0·28
Apples (with skin)	0·35
Bananas (with skin)	0·46
Green peas (boiled)	0·49
Milk	0·67
Cod (steamed)	0·81
Potatoes (boiled)	0·81
Eggs	1·62
Herring	2·36
Bread (white)	2·43
Roast beef (lean + fat)	3·21
Cheese	4·23
Bacon (fried, streaky)	5·26
Butter	7·97
Margarine	7·97
Beef fat	9·24

TABLE 5.17 Number of people whose requirement for protein or calories could be met from one hectare of the crops shown. (After Borgstrom, 1969)

	Soybean	Wheat	Corn	Rice	Potato	Sugar beet
Calories	5	8·4	10·4	14	16·5	34
Protein	14	6·3	5·2	7	9·5	12

from our different capacity to digest fibre, and prefer a more concentrated diet. Table 5.16 shows the relative concentrations of energy for various foodstuffs, to illustrate the relative quantities that would be needed to satisfy any given requirement. Another way of looking at this is to consider the number of people whose protein and energy requirements could be met from 1 ha of different crops (see Table 5.17).

Agriculturally, there are advantages in concentration of nutrients if the product has to be stored or transported. One of the most obvious aspects of concentration is a reduction in water content but this can be achieved in other ways, notably by drying. The latter requires substantial additional energy inputs and the cost is chiefly related to whether current or stored solar energy is used (see Chapter 7).

5. Elimination of Toxic Materials

Animals may achieve this in one of three main ways. First, they may feed selectively, more selectively than herbage could be harvested by cutting, for example, and thus reject material with poisonous or unpleasant properties. Secondly, they may excrete (or secrete) toxic materials, or even detoxify them. Thirdly, they may deposit such substances in parts of the body that are not usually consumed by man, including hair and bone.

6. Continuity of Food Supply

Since animals cannot live without food they cannot supply food to man at times when plant material would not have been available, although it may be possible to store animal products more easily or more efficiently: eggs are perhaps the best example, but carcasses can be readily stored in very cold climates.

What animals *can* do is to eat more food than they require at any one time and deposit reserves of fat: many animals have special depots for such storage. Thus animals can even out their supply of nutrients by storing food surpluses as fat for use when current intake is inadequate.

Furthermore, from our point of view, food used for animal growth also produces a living store and the slaughtering of animals can be spread over quite different periods from those involved in primary crop or herbage growth.

Cold-blooded animals can maintain themselves at very low cost during periods of low temperature and some warm-blooded animals can do the same (by hibernation), though the latter does not appear to have been exploited agriculturally.

Animal Production Processes

The ways in which animals fulfil their various roles vary greatly between species and environments but they are based on some six main processes:

1. Activity
2. Growth and Development
3. Reproduction
4. Secretion
5. Senescence
6. Death

They are not all strictly animal production processes but they are biological processes that underlie production from animal populations.

1. Activity

Under this heading may be included a number of behavioural characteristics of the animal, such as the way it feeds, its behaviour in relation to others of the same species, its behaviour in relation to man, its patterns of excretion (in space and time), and the food cost of all these activities. There are some behaviour patterns, such as the way in which herbivores defend themselves against attack by wild predators, that may not be required in an agricultural context, in which protection is supplied by fences or houses. These activities are therefore of rather less consequence, although it would be rash to assume that they are totally unimportant.

The major agricultural activities, of feeding, digesting, excreting and resting, are vital to the production process and bear a large part of the total food cost.

If an animal only eats enough to maintain its body weight, *all* the food consumed is "wasted" in these activities, since no production occurs. This "maintenance" requirement is very important in animal production, especially from warm-blooded animals. In the latter, it is generally related to some power of the live weight (W), varying between 0.5 and 1.0, but for mature mammals "resting" maintenance need appears to be approximately proportional to $W^{0.75}$ (Brody, 1945; Kleiber, 1947; Colburn and Evans, 1968). This expression is referred to as "metabolic size". Of course, actual need is related to the degree of activity and in homeotherms this is relatively independent of environmental temperature. In poikilotherms, on the other hand, activity and metabolism are greatly influenced by ambient temperature and their maintenance requirements may therefore exhibit very wide variations.

If animals do not even eat enough to maintain body weight, the food is not only unproductive but fat reserves are also used as a source of energy; in extreme cases, muscle is also drawn upon. In fact the mature cow, for example, is able to store and lose over 15 kg of body proteins (Paquay *et al.*, 1972).

The use of fat can be looked at in two quite different ways and both need to be taken into account. First, the use of fat for maintenance is very inefficient energetically. On a diet containing a high proportion of starch, for example, the energetic efficiency of fat deposition is about 64% (Blaxter, 1962).

Secondly, fat may be regarded as an alternative to storing food some other way (also not without losses), or not having any food at all, or death as soon as total food intake became less than that required for maintenance. As mentioned earlier in this chapter, one possible role of the animal involves the use of fat as a reserve, to cover discontinuities in the supply of plant food.

In natural conditions, fat is frequently used as an energy reserve and this might be considered to be one of its main functions (see Table 5.18 for examples of energy storage by animals). Furthermore, the whole notion of "maintenance-only" diets being rather wasteful only applies to some phases of animal reproduction, notably to the growth of homeotherms. For reproductive populations, for example, maintenance is exactly what is wanted for adult breeding animals over substantial periods and, for the male, for most of the time. This does not mean, of course, that the size of the maintenance requirement is unimportant. On the contrary, the higher the proportion of total food consumption that is used for maintenance, the more important is the magnitude of the maintenance requirement per unit of live weight. Variation in the latter is considerable (see Table 5.19), depending on species, breed, size, shape and degree of insulation afforded by coat cover or by subcutaneous fat.

TABLE 5.18 Fat reserves of animals
(from (1) Mason, 1967; (2) Talbot *et al.*, 1965; (3) Blake, 1967; (4) Epstein, 1969)

Species	*Site and nature of fat reserve*
Anatolian fat-tailed sheep (Red Karaman) (1)	Tail: 44 cm long x 31 cm broad; weight about 6 kg but up to 12 kg; represents up to 25% of carcass weight
Wildebeest (*Connochaetes* spp.) and wild herbivores in Africa generally (2)	Relatively constant, very low fat content (2–5% in the carcass)
Domestic cattle (mature beef animal: carcass weight 189 kg) (3)	Fat = 34·5% of carcass, distributed as follows: subcutaneous = 10·7% intermuscular = 17·7% kidney = 6·1%
Sheep (adult female: carcass weight 32 kg) (3)	Fat = 47% of carcass, distributed as follows: subcutaneous = 24·5% intermuscular = 16·5% kidney = 5·8%
Bactrian (two-humped) camel (live weight = 460 kg) (4)	Fat in both humps = 100 kg i.e. 21·7% of live weight
Reindeer (buck of 130 kg live weight) (4)	Fat in two dorsal areas over rump = 30–35 kg i.e. *ca.* 25% of live weight

There is some danger in referring to "maintenance requirement" as if it had one obvious meaning. In fact, it is most commonly expressed as energy, but the diet must always contain a minimum of certain minerals, nitrogen, vitamins and water, and there are large differences in the needs of different animals for these constituents. Nitrogen can be in an inorganic form for ruminants but not for pigs; so a herbage containing x% of total N has a different value for different animals, depending on the proportion of non-protein nitrogen in the total.

Animals differ in their ability to synthesize vitamins: ruminants, for example, because of the activities of their rumen flora, are independent of a dietary source of B vitamins.

In many ways, the most spectacular variation occurs in water requirement (see Tables 5.20 and 5.21), some animals requiring no free water in their diet and others having to drink large quantities every day.

For animals that are kept under controlled-feeding conditions, such as many pigs and poultry, the activity involved in food collection is not very important.

TABLE 5.19 Variation in maintenance requirement of adult animals. Maintenance requirement (M) is expressed per kg of body weight per day. (Sources: ARC; 1963; 1965; 1966; Large, 1974)

	M	
	g D.M.	kcal (gross energy)
Cow	7–13	27–57
Sheep	8–16	34–69
Pig	10–12	40–50
Rabbit	20	90
Hen	32	140–160

TABLE 5.20 Water turnover rate in animals (at 37°C) (Source: French, 1970)

Species	Water (ml per $kg^{0.82}$ per 24 h)
Camels	185
Goats	188
Sheep	197
Cattle	347

TABLE 5.21 Water requirements of the horse (source: Olsson, 1969)

Condition of the horse	Water requirement (kg/day)
Pregnant	40
Prior to foaling	42·7
Early lactation	57
Fattening	25–27
Resting	37
Working	68–86 (depending on rate of working)

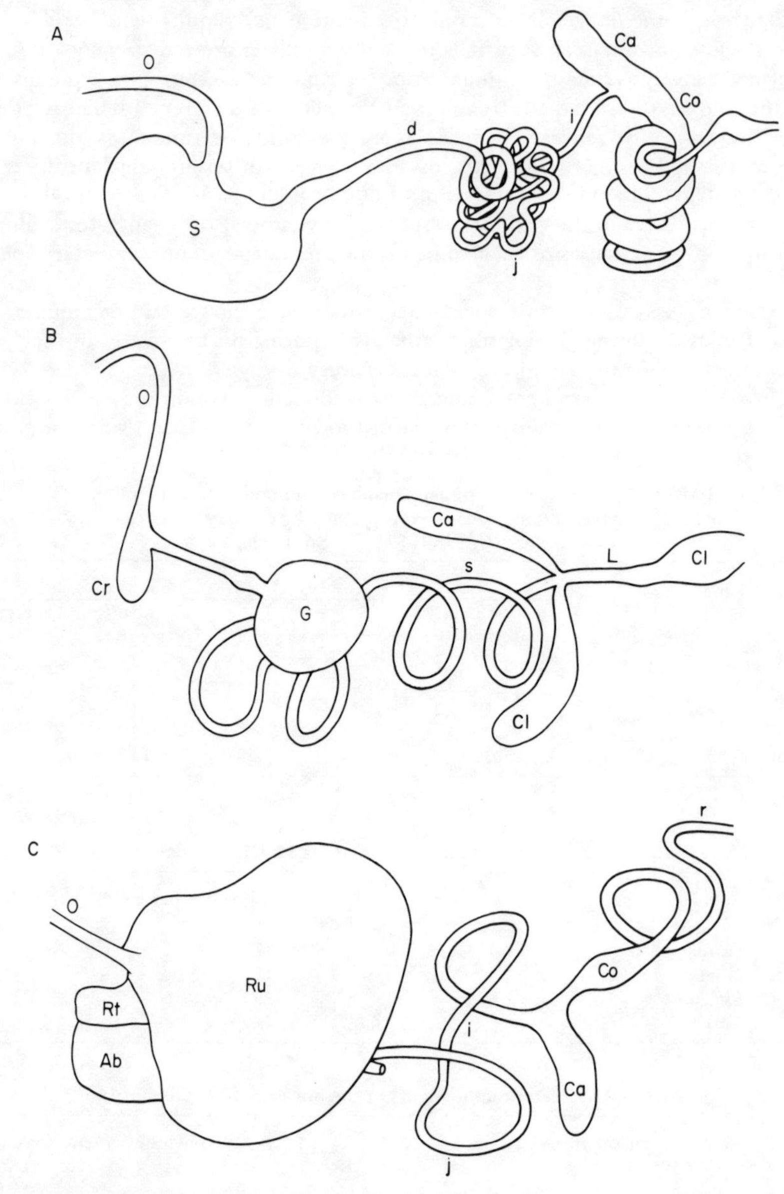

FIGURE 5.1 Diagrammatic representation of the digestive tracts of different farm animals: A, the pig; B, the fowl; C, the cow.

O	oesophagus	Ca	Caecum	Cr	crop	Cl	cloaca
S	stomach	Co	Colon	G	gizzard	Ru	rumen
d	duodenum	r	rectum	s	small intestine	Rt	reticulum
j	jejunum	i	ileum	L	large intestine	Ab	abomasum

For those animals that collect their own food, notably those that graze and browse, this activity may affect their efficiency in utilizing a particular environment. Frequently they are adapted in form and structure, both internally (as illustrated by the way in which they store and digest herbage) and externally (as exemplified by the length of leg and the nature of the foot).

The internal adaptation has already been mentioned and is illustrated in Fig. 5.1.

External structures concerned with method of feeding include the mouth, with a horny pad instead of upper incisors in most ruminants, with continuously-growing incisors in rabbits and molars in elephants. Lips may be well-developed and mobile for seizing herbage (as in horses), or divided to get them out of the way (as with the upper lip in sheep). Tongues may be long, to help in pulling off herbage (as in cattle) and are generally tough.

If legs are long, the neck has to be long also, unless a trunk is present as with elephants and, to a lesser extent, tapirs.

Sensory organs (ears, nose and eyes) are well-developed in grazers, since these animals are very susceptible to predation and, for speed, legs are usually long, the feet make little contact with the ground and are often protected by renewable hooves. The degree of contact with the ground also influences the extent and the way in which pasture may be damaged by grazing animals, so the pressure per unit of hoof area is of some interest (see Table 5.22).

There are some marked exceptions to the foregoing, wherever animals are adapted to a particular terrain. The camel combines large feet, for use on loose sand, with speed, although it is in any case a large animal with few enemies; the buffalo is adapted to wet land and does not rely on speed at all, except perhaps over extremely short distances.

It is difficult to interpret such attributes as length of leg and it is probably a gross oversimplification to consider such features in isolation: the leg length of different grazing animals has to be related to the absolute size and weight of the animal considered, for example. Figure 5.2 shows the relationship between the ratio of height at withers (because this is better defined than leg length) and body weight.

TABLE 5.22 Hoof area and pressures, measured on individual animals (the range of weight given represents the annual variation in the live weight of the individual) (From Spedding, 1971)

	Total hoof area (cm^2/animal) (a)	Weight of animal (kg) (b)	Pressure (g/cm^2) (b/a)
Sheep			
(Kerry Hill)	92·4	74·3–87·5	800–950
	79·8	59·0–73·5	740–920
Cattle			
(South Devon)	350	500–560	1430–1600
(Jersey)	250	320–365	1280–1460

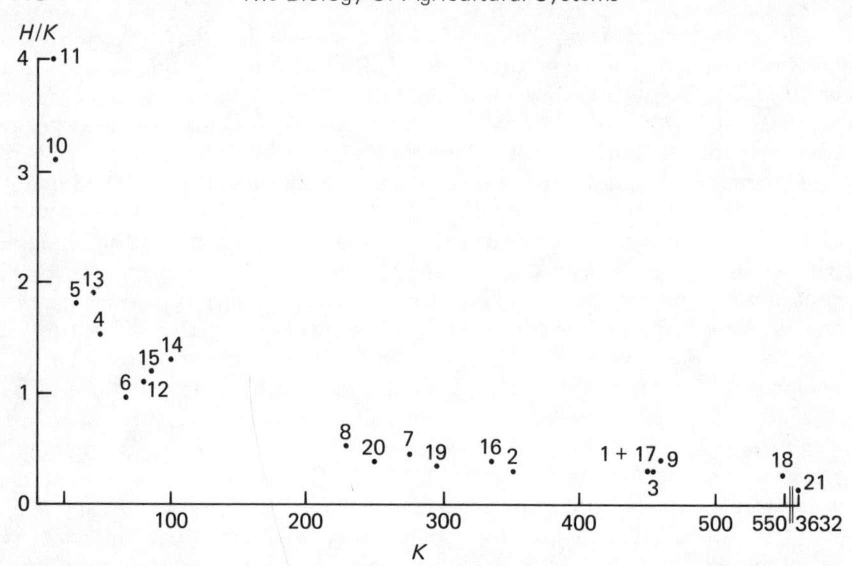

FIGURE 5.2 Height/weight ratios in animals. Height (*H*) refers to height of withers (in cm) and weight (*K*) is in kg: the ratio has been calculated for the following animals:

1. Bull of Humpless Cattle of N.E. China (Manchuria)[a]
2. Yak Bull[a]
3. Chinese Buffalo[a]
4. Hu Yang ram[a]
5. S. China goat (buck)[a]
6. Tungcheng pig (boar)[a]
7. Mongolian pony[a]
8. Suimi Ass (♂)[a]
9. Dromedary (1 hump ♂)[a]
10. S. China pariah dog[a]
11. Chinese greyhound[a]
12. Alpaca[b]
13. Vicuña[b]
14. Llama[b]
15. Huanaco[b]
16. Ane (donkey)[c]
17. Mulet (mule)[c]
18. Bardot (Hinny)[c]
19. Yak ♂[d]
20. Yak ♀[d]
21. Elephant (adult)[e])

Such a ratio is clearly unsatisfactory for specialized animals (e.g. Rabbits and Kangaroos).

[a] Epstein (1969)
[b] French (1966)
[c] Bonadonna (1966)
[d] Schulthess (1967)
[e] Petrides and Swank (1965)

Attributes of this kind may be of considerable relevance in those agricultural animals that are used for transport or traction.

Only mammals have been used to any extent as beasts of burden, partly because of their size and partly because of the plasticity of their behaviour—they can be tamed and trained fairly easily (Parker, 1947). Apart from the dog and the elephant, furthermore, beasts of burden have mostly been hoofed mammals. The dog has been most used in snow-covered regions, where weight is a disadvantage, although the reindeer is also much used in similar regions. At the other end of the adaptation range, the camel is outstanding in the desert. There are many other animals, however, adapted to particular areas, that have been used for a very long time. Table 5.23 shows the main animals of the world that have been used for transport, traction or in battle.

TABLE 5.23 Animals and their "non-product" uses
(Sources include: Turner, 1971; Parker, 1947)

	Transport	Draught	Battle
Ass	×	×	
Buffalo	×	×	
Cattle	×	×	rarely
Camel	×	×	×
Dog	×		
Donkey	×		
Elephant	×	×	×
Horse	×	×	×
Llama	×		
Mule	×	×	×
Reindeer	×	×	
Sheep	×		
Yak	×		

2. Growth and Development

Although "growth" is an extremely useful term and a very important biological concept, it is not easy to define satisfactorily.

It involves the notion of increase but sometimes in one dimension and sometimes another. Clearly, a leg can grow in length, even if it then weighs less; equally, growth can be represented by weight increase, even if no increase in length or volume occurs. Increase in total live body weight is a very commonly used index of animal growth and has then to take account of water in the coat and food in the alimentary tract. But weight loss can occur in an animal that has grown larger skeletally, so it is clearly necessary to specify what kind of growth we are talking about.

Development is often involved also, and growth might be implied as much by an increase in cell number as by an increase in cell size. It is for this kind of reason that the linked term "growth and development" is often used.

Growth and development often occur together in animals and this is what is involved in the rearing and growing of agricultural animals for breeding or for meat production.

It may be useful sometimes to think of fat deposition as something different, as in the case of weight changes in adult animals simply due to changes in depot fat. Nevertheless, fat deposition certainly occurs during growth and is part of development. The agricultural distinction between "growing" and "fattening" does not usually relate to the deposition of depot fat so much as to the later stages of growth when a high proportion of the weight gained is sub-cutaneous or inter-muscular fat.

It will be clear that live weight gain only represents growth in a crude fashion, since the weight gained may be bone, muscle, fat or water (held in the tissues). Comparisons of growth rate are fraught with difficulties of this kind, not only

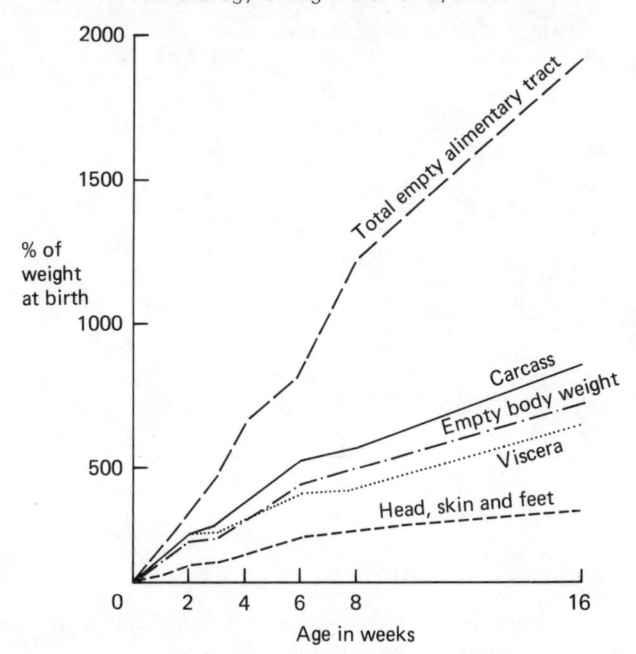

FIGURE 5.3 Relative changes in body composition of the lamb. Source: Spedding (1970), based on data from R. V. Large.

between species or individuals, but within the life of one animal because its body composition changes markedly as it grows (see Fig. 5.3).

Agriculturally, however, growth rate is extremely important and comparisons have to be made. Growth curves are usually considered to be sigmoid in form and this appears to be generally so, independent of the methods of expression that seem most relevant (see Fig. 5.4), provided that the whole period of growth is considered. Over portions of this period, the growth curve may often be linear.

Although growth rate is related to parental size, different species of animals do not reach maturity at the same age. However, there is some tendency for larger mammals to live longer as well as to grow faster and have a longer gestation period: if the times for growth and gestation are calculated as a proportion of the average life span, they do not differ so greatly (see Table 5.24). Taylor (1965) has found that "the time a species takes to reach any particular degree of maturity, that is any fraction of its mature weight, tends to be directly proportional to its mature weight raised to the 0.27th power". Thus, "metabolic age" (θ) may be defined as:

$$\theta = \gamma(t - t_0)A^{-0.27}$$

where t = age from conception

t_0 = an age origin at or near conception

γ = a constant, conveniently = to 1

A = mature live weight.

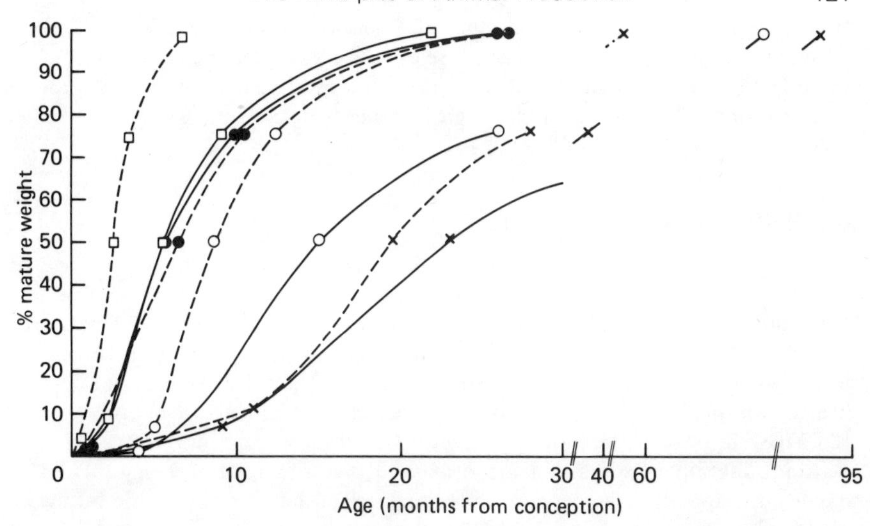

FIGURE 5.4 Growth curves of animals, expressed in terms of live-weight as a % of mature weight related to age in months from conception, for:

Cattle	x ——————— x	Rabbits	● ——————— ●
Horses	x — — — — — x	Fowl	● — — — — — ●
Swine	○ ——————— ○	Guinea pigs	□ ——————— □
Sheep	○ — — — — — ○	Pigeons	□ — — — — — □

(Source: Brody, 1945; Fishwick, 1953; Donald and Russell, 1970; Vaccaro *et al.*, 1968.

TABLE 5.24 Length of time to first breeding and for gestation (or incubation) as % of life span (in agricultural practice)

	A Age at first breeding[a] (days)	B Mean gestation length or incubation period (days)	C Average life span (days)	$\frac{A}{C} \times 100$	$\frac{B}{C} \times 100$
Cattle	730 (1)	280 (2)	4015 (3)	18·2	7·0
Sheep	730 (1)	147 (2)	2190 (3)	33·3	6·7
Rabbits	120–150 (1)	31 (2)	730 (1)	16·4–20·5	4·2
Horses	912–1095 (5)	336 (2)	6935 (3)	13·2–15·8	4·8
Pigs	310 (3)	115 (2)	1095 (4)	28·3	10·5
Hens	168 (1)	21 (8)	548 (4)	30·7	3·8
Geese	273–365 (6)	30 (8)	1643 (6)	16·6–22·2	1·8
Turkeys	212 (7)	28 (8)	365 (7)	58·0	7·7

[a] birth to first parity for mammals, hatching to first egg laying for birds.

Sources:
1. Large (1973)
2. Kenneth and Ritchie (1953)
3. Fraser (1971)
4. Wilson (1968)
5. Asdell (1964)
6. Ontario Dept. of Agriculture and Food
7. Walsingham (1974)
8. Kosin (1962)

In fact, apart from the special case of gestation, there are even similarities between mammals and birds; in the case of cold-blooded animals growth is more variable and, in some cases (e.g. fish), there is no obvious mature size and growth may continue throughout life. There is, nevertheless, an age or size when breeding first becomes possible.

3. Reproduction

In agricultural animals reproduction is either by the production of eggs or offspring and generally involves both male and female individuals. The few exceptions are animals like snails, which are hermaphrodite. Since the male may only be required for purposes of insemination and is often large, with a food intake to match, the ratio of males to females (see Table 5.25) has a considerable bearing on the food conversion efficiency of a population; artificial insemination can have a dramatic effect on the number of males needed, however.

Reproduction in some animals is markedly seasonal, often being related to day-length, and the existence of a breeding season has a considerable influence on breeding frequency (see Table 5.26); gestation length is another factor of importance.

TABLE 5.25 "Normal" performance levels and attributes of animals
(Information derived from the literature)

Species	Mature size live weight (kg) Male	Mature size live weight (kg) Female	Reproductive rate No. of young per year	Ratio of males to females for breeding	Yield/ progeny carcass weight (kg)
Cattle	700–800	450–700	0·9	1 : 30–50	200–300
Buffalo	665–718	509–548			144–279
Musk ox	365		0·5–1·0		
Yak	230–360	180–320			
Sheep	30–150	20–100	1–2+	1 : 30–40	18–24
Goat	48–58	45–54	1–3	1 : 40	4·3–8·4
Deer (red)	124	75–82	1	1 : 1·6–6·6	20–64
Horse	1000	700–900	1	1 : 70–100	360
Camel	450–840	595	0·5	1 : 10–70	210–250
Alpaca	80				
Llama	80–110				
Rabbit	4·0–7·2	4·5–7·6	30–50	1 : 15–20	1–2
Guinea-pig		0·6–1·0	20–30		0·7
Capybara	60	45	8·7		15·3
Pig	350	220	20	1 : 20	45–67
Dog	12–15				
Hen	4	3	108	1 : 10	1·45
Duck	4·5	4·0	110–175	1 : 5–8	2
Goose	5–10	4·5–9	25–50	1 : 2–6	4–5
Turkey	13–23	8–12	40–100	1 : 10–15	3–9

(from Spedding and Hoxey, 1974)

TABLE 5.26 Breeding season of female animals

	Duration of oestrus	Oestrus cycle (days)	Gestation or incubation (days)	Breeding	
				Season[a]	Life (years)
Cattle	18 h (1)	21 (1)	280 (1)	All year (1)	8–14 (1)
Sheep	26 h (1)	16·5 (1)	147 (1)	Mainly autumn to spring (1)	6 (1)
Goats	28 h (1)	19 (1)	150 (1)	Mainly autumn to winter (1)	6–10 (1)
Rabbits	Inducible (1)	– (1)	31 (1)	All year (1)	2 (5)
Horses	6 days (1)	21 (1)	336 (1)	Spring and summer (1)	16–22 (1)
Pigs	2 days (1)	21 (1)	115 (1)	All year (1)	3 (6)
Buffalo	12 h (2)	21 (2)	310 (2)	All year (2)	Up to 20+ (3)
Hens	–	–	21 (4)	All year (4)	$1\frac{1}{2}$ (6)
Turkeys	–	–	28 (4)	Spring (4)	1 (7)
Geese	–	–	30 (4)	Spring (4)	4–5 (8)
Guinea fowl	–	–	28 (4)	Spring (4)	?

[a] Breeding seasons tend to be less marked in the tropics

Sources:
1. Fraser (1971)
2. Asdell (1964)
3. Cockrill (1967)
4. Kosin (1962)
5. Large (1973)
6. Wilson (1968)
7. Walsingham (1974)
8. Ontario Dept. of Agriculture and Food

Egg production serves as a basis for rearing breeding stock, as a basis for meat production and directly as a production process. Newly-born offspring very rarely serve as a product, perhaps because of their relatively high production cost, unless they are exceptionally valuable. Some very young pigs and lambs are slaughtered (at a few weeks old) for special dishes (often of religious or ceremonial significance) and the production of astrakhan is based on the slaughter of newly born Karakul lambs, before the curl of the coat has decreased.

Egg production has a number of special characteristics: it can give a daily product that is sterile, storable and of high nutritive value; the eggshell is a convenient, though fragile, container; and fertilization is not required where eggs are solely for eating. In relation to the last point, in some bird species a male has to be present before eggs will be laid and, in the case of pair-bonding birds, an equal number of males and females is needed.

In general, of course, it is an advantage in agricultural animals to minimize the total size and food requirement of the breeding population per unit of product produced. For any given parental unit, therefore, there are advantages in a high reproductive rate. The latter has to be related to so many factors that there is some merit in expressing it as the number of progeny per annum per unit weight

TABLE 5.27 Number of progeny per parental unit

Animal	Male (a) : female (b) ratio*	No. of progeny/♀/year[a] (c)	$\frac{c}{1 + a/b}$
Cattle (using A.I.)	1 : 2000	1·0	0·9995
Sheep	1 : 40	1·5	1·46
Rabbits	1 : 20	40	38·1
Hen	1 : 15	240	225
Goose	1 : 4	40	32·0

[a] These vary but typical values have been selected for the purposes of this table.

of parental unit (i.e. males and females necessary to produce the progeny). This has been calculated (Table 5.27) for a range of agricultural animals.

In many ways it would be even better to calculate the proportion of the total food used by a population that is devoted to reproduction (see also Chapter 4). In both cases, the extent to which any given reproductive rates result in additional mature adults depends upon a host of environmental factors. It might be expected that high reproductive rates would have evolved in circumstances where these "environmental" losses were high and that agriculture, with its greater control of the environment, would be able to exploit animals of this kind. To some extent this is true but there are, of course, many other factors (e.g. mature size) that govern the choice of agricultural animals.

One major difference between reproducing animal species is the degree of involvement of the parents after "birth". Eggs are often wholly independent starting-points for the growth process but, in birds, they require warmth during a substantial incubation period. In fish and reptiles, no attention may be needed at all. Mammalian offspring require a great deal of attention and rely nutritionally on their dams for considerable periods.

It is important to recognize that much agricultural development has depended upon substituting artificial procedures for these natural relationships. Most eggs, in developed agricultural systems, are incubated artificially and chicks never see their parents. Modern dairy production is based on the removal of calves, virtually at birth, to be reared on milk-substitutes, often on a different farm.

Pig and sheep farming could go the same way, in that it is already technically possible, but up to the present time have not done so to any great extent. Early-weaning techniques have reduced the period during which suckling occurs, however.

4. Secretion

By far the most important secretion from an agricultural animal is milk. Others include semen for use in artificial insemination, which may be small in volume but of considerable monetary value, and silk from larvae of the silk moth

(*Bombyx mori*). There are further possibilities for the future, such as mammalian eggs for implantation in other females of the same species, and there are some secretions of indirect importance, such as the wax and "suint" that forms part of the fleece in wool production. It is doubtful whether the "milking" of poisonous snakes for the production of antidotes can be classed as agricultural, although logically it must be regarded as animal production.

Milk and silk appear, therefore, to be the main secretions of importance.

a. Lactation

Milk for direct human consumption is produced mainly by cows, goats, buffalo and sheep (see Table 5.2), although it may first be converted into butter or cheese (partly for preference and partly for storage).

The composition of milk varies from one species to another (Linzell, 1972) and also varies enormously in absolute yield. If lactation curves are considered in relation to length and peak yield, however, they tend to be of similar shape. The range of milk yields for the most relevant animals is illustrated in Tables 5.28 and 5.29.

The quantity secreted depends greatly on the rate of removal and clearly cannot exceed it. This means that frequency of milking matters in dairying and, in suckling, frequency, number of progeny and amount removed each time suckling occurs, are all important. Suckling frequency varies between species but generally declines with time (Table 5.30).

In general, species that suckle frequently, on demand, tend to produce milk with lower concentrations of nutrients (Ben Shaul, 1962). All the species in agricultural use for milk production are in this category. Species that suckle

TABLE 5.28 Milk yields of animals (after Linzell, 1972)

Animal	Representative milk yield/day (kg)	Representative peak milk yield/day (kg)	Animal weight (kg)	Weight$^{0.75}$	Representative milk yield (kg per day per kg$^{0.75}$)	Highest recorded milk yield (kg per day per kg$^{0.75}$)
Guinea pig	0·070	0·093	0·833	0·873	0·08	0·1065
Rabbit (NZW)	0·285	0·420	4·5	3·09	0·0922	0·1359
(Dutch)	0·140	0·166	2·4 1·77	1·93 1·521	0·0725	0·1091
Sheep (Suffolk, Hampshire)	2·3	3·7	70	24·2	0·095	0·1528
Goat (Saanen)	5·0	10·0	70	24·2	0·2066	0·4132
Pig	10·5	12·0	204 153	54 43·9	0·1944	0·2733
Horse (pony)	7		277	68·4	0·1023	
(draught)	14		580	118·2	0·1184	
Beef cow	11	16·1	500	105·7	0·104	0·1523
Friesian cow	43	90	550	113·6	0·3785	0·7922

TABLE 5.29 Peak milk yield per unit of metabolic body size (kg per day per $kg^{0.75}$) (After Linzell, 1972)

	Peak milk yield	
	Representative	Highest recorded
Guinea pig	0·08	0·1
Rabbit	0·07 0·09	0·11 (Dutch) 0·14 (NZW)
Sheep (Suffolk, Hampshire)	0·1	0·15
Goat (Saanen)	0·2	0·4
Pig	0·19	0·27
Horse (pony)	0·1	–
Beef cow	0·1	0·15
Friesian cow	0·4	0·8
Buffalo	0·11	–
Horse	0·12	–

TABLE 5.30 Suckling frequency in animals in relation to age of progeny (frequency also varies with, for example, litter size)

	Number of occasions per day (24 h)			
	Age (days)			
	0–10	10–20	20–40	40–80
Calf	6–8[2,4]	8[4]	3[3]–6[4]	5[4]
Lamb	40[5]–78[1]	30[5]	6[1]–20[5]	16[5]
Rabbit	1[6]	1[6]	–	–
Piglet	24[1]	20[1]	18[1]	16[1]

1. Hafez (1962)
2. Roy (1959)
3. Johnstone *et al.* (1964) (Age of calf not stated)
4. Ewbank (1969)
5. Ewbank (1967). (The figures in the table have been obtained by doubling those obtained in 12 h periods of observation)
6. Cross and Harris (1952).

infrequently, on a scheduled basis, tend to have much higher concentrations, especially of fat. If the volume of milk is low, the only way to provide enough energy for the young is by a high fat content. As Jenness (1972) has pointed out, for all the species of animals for which milk composition has been determined, the total energy contribution from the proteins and carbohydrates falls between 30 and 65 kcal per 100 g, whereas fat supplies additional energy up to 500 kcal per 100 g.

The secretion rate of a wide range of mammals falls close to 2 g milk per g mammary tissue per day (126 ± 17 g milk per $kg^{0 \cdot 75}$ of body weight/day).

Lactation, is, however, influenced by current nutrition, including availability of water, but also by previous nutrition, for two main reasons. First, there may be a direct effect towards the end of pregnancy on the amount of udder tissue formed. Secondly, if milk output exceeds the capacity of current food intake to support it, it may still continue by drawing upon fat reserves, provided that these are present at the beginning of lactation. This means that previous nutrition was either adequate for their formation or, at least, did not involve their depletion.

b. Silk Production

Silk is produced from the cocoons of certain moths, notably *Bombyx mori*, and in Japan, for example, the latter has been the basis of a substantial industry for centuries. Sericulture is practised in more than twenty countries, however, and several other species of silk-moth are used, as well as *Bombyx mori* (Yokoyama, 1963).

In Japan, silkworms are fed on mulberry leaves, cultivated for the purpose. There are ten species and at least 39 varieties of mulberry in Japan, *Morus alba* being the species most widely used. The object is to produce as many leaves as possible per unit area of land and triploid plants are used, liberally supplied with nitrogenous fertilizer.

There are more than 1000 commercial varieties of silkworm and much research has been done on both breeding and rearing.

Silk itself consists of two proteins, fibroin and sericin, the former making the fibre: it is secreted by both male and female larvae prior to pupation. The pupae, after the silk has been removed, may be used as a source of animal feed, especially for fish (Hickling, 1971).

In Japan, a unit ("one box") of silkworms is 20,000, derived from 10 g eggs: these larvae consume about 500-600 kg of mulberry leaves and produce some 30 kg of cocoons (Takeuchi, 1967).

Silkworms are unusual (in farming) in that fertile eggs can be stored at low temperatures for several months.

5. Senescence

Senescence does not have quite the same significance in animal production as it has in crop growth (see Chapter 4).

It is of little relevance to all those animals that are slaughtered before or at maturity but is of consequence to breeding stock and animals kept for the production of eggs, milk, wool and work.

The main significance lies in the rather obvious fact that rates of production and performance tend to fall beyond a certain age. This has the effect of shortening the productive life of an animal and its "economic" life span may be even shorter still. Most animal production systems therefore involve a culling system, with animals taken out as they fall below production or performance standards. These "culls" often contribute significantly to the total meat production, however.

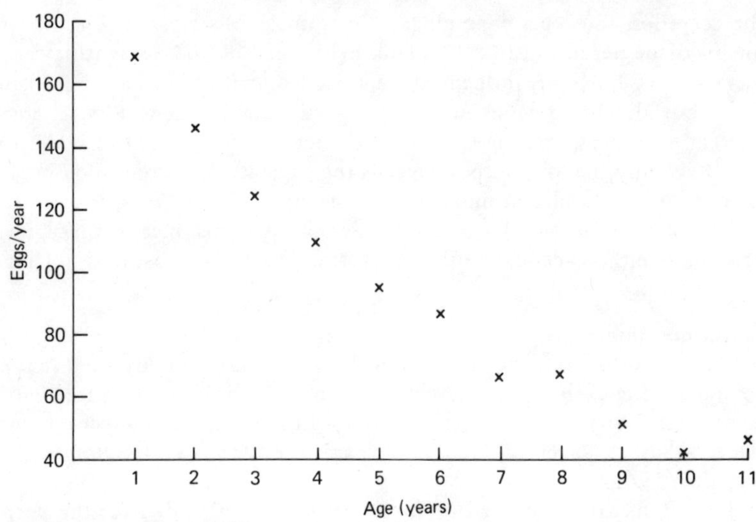

FIGURE 5.5 The effect of age on egg production in the hen (from Hall and Marble, 1931, quoted by Brody, 1945).

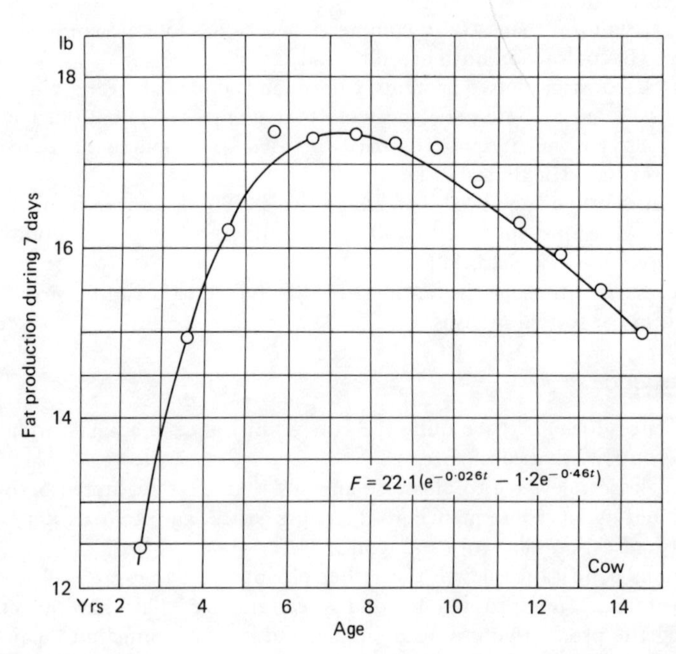

FIGURE 5.6 The rise and decline of milk yield with increasing age as indicated by 7-day records from the Advanced Register of Holstein cows. The circles are observed values; the smooth curve represents the equation in the chart (from Brody, 1945).

Common examples of falling productivity are egg production in poultry, prolificacy in sheep, capacity for work in horses and milk yield in cattle (see Figs 5.5 and 5.6).

6. Death

Death of the individual is an event rather than a process but may be regarded as a continuous series of events in a population.

The rate at which it occurs may be controlled, as in culling (see above) or direct slaughter for meat, or may be due to factors such as starvation, predation, poisoning, heat or cold stress, disease or accident (e.g. drowning).

Causes of mortality vary enormously with species, system and environment, but agricultural systems ought to provide adequate nutrition and protection from predators, heat and cold. Thus poisoning ought only to be accidental and failure in other respects might be similarly described. However, failure or the risk of failure may be due to lack of knowledge or of skill in applying it, or to the fact that it would be uneconomic to eliminate all such risks.

Disease will continue to be a major factor because its incidence is not static. Disease-producing organisms can change their nature or their ways and new ones can arise, just as agricultural systems change and alter the stresses to which the animal is subject.

Efficiency in Animal Production

"Efficiency" can bear many meanings but the most useful general definition involves a ratio of "output" per unit of "input". Even if this meaning can remain constant, however, it is clear that its *expression* must vary widely. There is no right or wrong ratio, therefore, only expressions that are more or less appropriate to the purposes for which they are employed (see Spedding, 1973).

If we wish to compare animals (or plants) for their efficiency at producing protein in the desert, it may be of great value to compare them in terms of the ratio of (nitrogen output)/(available water), when water supply may be the limiting factor to production and where nitrogen is a good index of protein. The last point is typical of the kind of reservations which are often held in relation to a given ratio, because so often interpretation of its meaning depends not so much upon the ratio itself as on what it is thought to reflect or indicate.

Thus, the ratio output/input exemplifies efficiency, whether a product is involved or not and provided that time is accepted as an input: biological efficiency implies a biological output from a biological process but the inputs are not so limited. Although such a wide definition appears to be most generally useful, it makes it imperative that the purpose of using a particular expression be clearly stated, so that the terms used can be justified as those most useful for that purpose.

Two further contexts have to be specified, relating to space and time.

The physical environment has to be stated, in which the process is being assessed, for obviously the efficiency of a process will depend to a very large

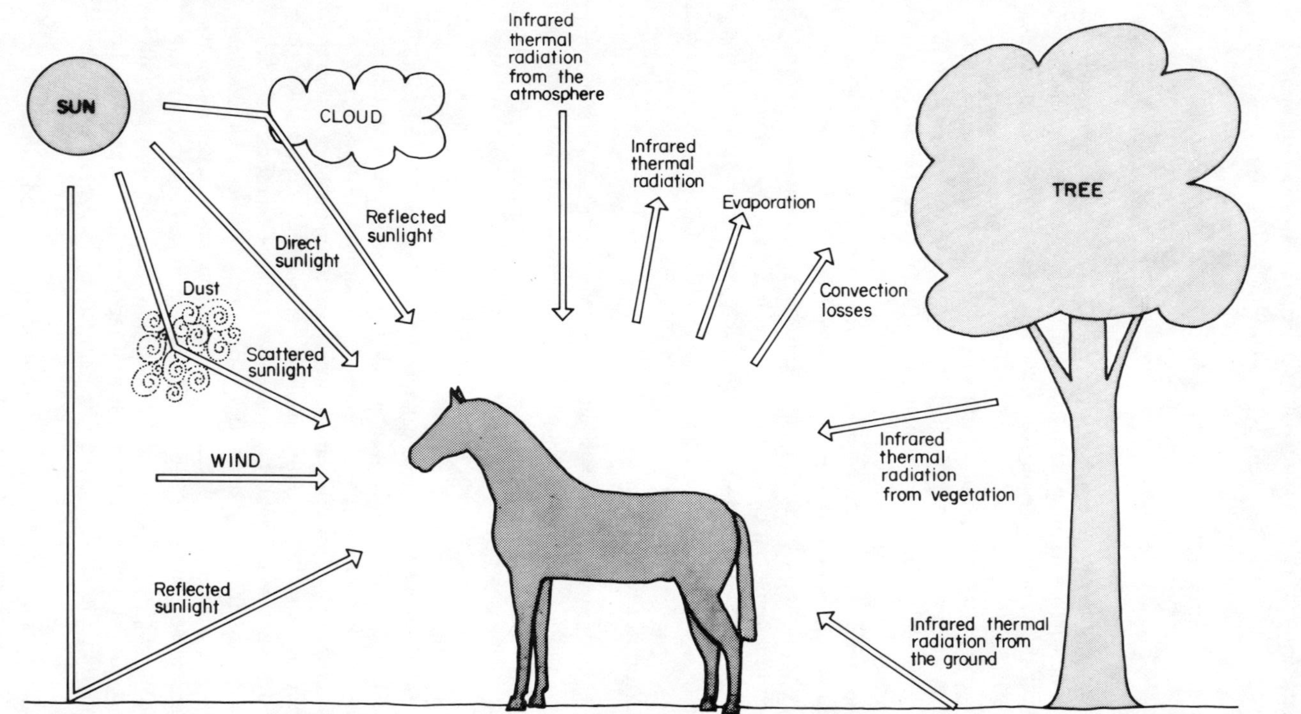

FIGURE 5.7 A diagram to illustrate the non-food stream of energy between an animal and the environment (after Porter and Gates, 1969). The energy budget can be calculated from:

$$M + Q_{abs} = \epsilon\sigma T_r^4 + b_c(T_r - T_a) + E_{ex} + E_{sw} \pm C \pm W$$

where ϵ = emissivity of the skin; σ = Stefan Boltzmann constant; M = metabolic rate; Q_{abs} = radiation absorbed by animal surface; b_c = convection coefft.; C = heat conducted to substrate; W = work done; T_a = air temperature; T_r = animal's temperature (surface); E_{ex} = respiratory moisture loss; E_{sw} = moisture loss by sweating.

extent on the environment in which it is carried out. A statement that pigs are more efficient than sea-lions at converting food to meat tells us little unless we define both food and meat and state whether we have a terrestrial or marine environment in mind. Furthermore, the appropriate description of the environment must depend upon the measures of efficiency that are being compared.

This does not mean that comparisons are only valid or useful *within* environments, simply that the environment must be stated and, indeed, adequately described. It is quite conceivable that one would wish to compare two animal or plant species when each was operating under its preferred environmental conditions, even if the latter were totally dissimilar. It has to be remembered, of course, that the choice of animal for a particular environment may have more to do with climatic conditions, such as rainfall, than with food as such (Ruthenberg, 1971). Figure 5.7 illustrates the non-food exchanges of energy between an animal and its environment.

Similarly, the period of time over which efficiency is measured is of great importance and must be stated. An adequate description of the period involves much more than a statement of time itself, however. A year, for example, may be fine for sheep production, taking into account a full cycle of herbage growth and allowing a complete breeding cycle for the sheep, but it may represent less than one breeding cycle for a cow, much less than one cycle for an elephant, several cyles for a rabbit, and several generations for mice and insects.

The food conversion efficiency of an individual animal changes with age, diet, size and season and the efficiencies of populations vary according to whether a lifetime or one reproductive cycle is used as the basis for the calculation (Spedding, 1971; Large, 1973).

Frequently, one complete breeding cycle is of interest but breeding frequency, differential mortality and longevity are then all ignored. Added to this, animal populations differ in the nature of their main products and by-products and in whether they are self-replacing or replenished from another source.

There are therefore many different periods that may be of interest and the only essentials are that they be precisely stated and chosen in relation to the purpose for which efficiency is being calculated.

It is not possible to discuss all the efficiency ratios that may usefully be calculated for all animal production processes, but it may be helpful to consider the expressions that are of most agricultural relevance for the main processes.

Expressions of Efficiency in Animal Production

The ratio of output per unit of input can generally be written as P/R where P = quantity of product or performance and R = the quantity of one or more resources.

The major elaborations of this ratio therefore concern different expressions (such as energy, protein, weight, money) for the main products and resources (see Table 5.31).

TABLE 5.31 Formulae used for the calculation of efficiency in animal production

Formula	Reference
$E^{(1)} = \dfrac{P}{\dfrac{M}{K_m} + \dfrac{milk}{K_1} + \dfrac{eggs}{K_e} + \dfrac{wool}{K_w} + \dfrac{total\ gain}{K_f} - \dfrac{total\ loss}{K_m}}$	Holmes (1971)
$E^{(2)} = \dfrac{N \times C}{D_p + D_1 + D + (N \times Y)}$	Spedding (1971)
$\dfrac{Expense/yr.^{(3)}}{Product/yr.} = \dfrac{\overbrace{(A/Y) + (I_d + B_d \cdot \bar{F}_{md} + F_{pd})}^{\text{For breeding female}} + \overbrace{N[D(I_o + B_o \cdot F_{mo} + F_{po}) + S_o]}^{\text{For her progeny}}}{P_d \cdot V_d + N \cdot P_o \cdot V_o}$	Dickerson (1970)
$\text{Gross Efficiency}^{(4)} = \dfrac{\text{Output energy (of milk, eggs, meat, work etc.)}}{\text{Input energy (of gross energy, T.D.N. etc.)}}$	Brody (1945)
$E_B{}^{(5)} = \dfrac{\dfrac{12}{i}\left(p - \dfrac{100}{d}\right) W_L \cdot Q_{c_L} + \left(\dfrac{100}{d} - v\right) W_s \cdot Q_{c_s}}{F_{Me} \cdot Q_{F_{Me}} + F_{ML} \cdot Q_{F_{ML}} + F_Z \cdot Q_{F_Z} + F_L \cdot Q_{F_L} + F_B \cdot Q_{F_B}}$	Wassmuth (1972)

(1) Generalized equation for overall efficiency of animal production:

P = product as energy or protein

K_m, K_1, K_w, K_e, K_f are decimals expressing efficiency of nutrient use for maintenance, lactation, wool production, egg production, fattening

(2) For meat production of dam/progeny units:

D_p = food required by the dam during pregnancy
D_l = food required by the dam during lactation
D = food required by the dam during the non-pregnant, non-lactating period
Y = food required by each of the progeny
N = number of progeny
C = carcass weight of each progeny

(3) A/Y = (Cost, young female – value, old female)/years in production
I_d = Yearly fixed costs/♀
B_d = Metabolic body size of ♀, relative to population mean
$\bar{F}_{md}$ = Average in maintenance feed costs/♀/year
F_{pd} = Feed costs *above* maintenance/♀/year
N = Number of progeny marketed/♀/year
D = Days from weaning to marketing
I_o = Average fixed costs/animal/day
B_o = Average post-weaning metabolic body size for individual, relative to population mean
$\bar{F}_{mo}$ = Average maintenance feed cost/animal/day
F_{po} = Average feed costs above maintenance/day/individual
S_o = Fixed costs/animal for slaughter, marketing, vaccines et cetera
P_d = Annual volume of product/♀
P_o = Live weight of meat animal when marketed
V_d = Value per unit of ♀ product
V_o = Value per unit of live weight

(4) Brody (1945) distinguished "gross" efficiency, which included energy costs of maintenance, from "net" efficiency, which excluded maintenance

(5) Biological efficiency (E_B) is expressed in terms of:
i = interval between lambings
p = Lambing %
d = Average productive life of the ewe
W_L = Weight of lamb
W_s = Weight of cull ewes
V = Losses of lambs and ewes
Q_c = Carcass quality
F_{Me} = Maintenance requirement of ewes
F_{ML} = Quantity of food for production
F_Z = Food for replacements
F_L = Food for lambs
F_B = Food for rams
Q_F = Food quality (expressed as starch equivalent)

Thus the efficiency of egg production per unit of food consumed could be expressed as:

$$\frac{\text{weight (or energy content, or number, etc.) of eggs produced}}{\text{weight (or energy content or nitrogen, etc.) of food consumed}}$$

over a stated period of time and in a given environment. But this does not specify the animal sufficiently and it is essential to define the animal population involved (for we rarely intend the calculation to apply only to one unique animal).

The animal populations used for animal production vary greatly but they do embody a common structural pattern (illustrated in Fig. 5.8) and efficiency

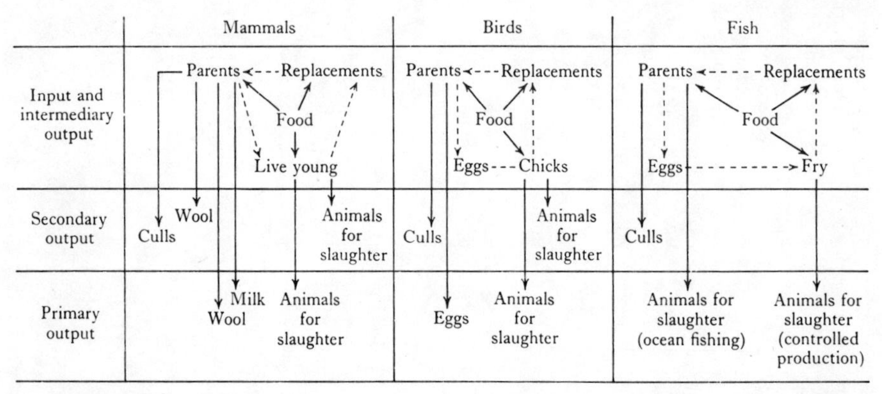

FIGURE 5.8 The structure and outputs of animal populations. (N.B. A basic 'wastage' rate from disease and mortality may occur at any stage; in the interests of simplicity this is now shown.) (From Large, 1973.)

must refer to some specified part of this. Not all animal species actually exhibit all the phases in the pattern and not all parts of the process are normally used as products, however; furthermore, quite small parts of the total process for any one species may be the subject of an efficiency study. These facets are also indicated in Fig. 5.8.

Three kinds of population are worth special emphasis.

1. Individual Progeny

Groups of progeny (e.g. calves or chicks) often form the basis of a meat-production enterprise. An economic ratio will take account of their initial cost but calculations of biological efficiency normally do not (or, rather, if they do, they are not strictly concerned only with the individuals themselves). There is no reason why such calculations should not be made, and indeed they are highly pertinent to some enterprises, but it is misleading, of course, if such ratios are misused, to represent, for example, the relative efficiency of different animal species.

The efficiency of an individual generally relates only to those individuals of which it is a fair sample. However, where the cost of producing an individual is so low as to be negligible, the individual may even be a fair sample of the whole population.

In the mayflies (order: Ephemeroptera), for example, the adult flies live for a matter of hours or days and consume no food at all, their mouth-parts being aborted. The food cost per egg is therefore nil. Yet the larvae live for substantial periods of time (up to a year or more) and consume large quantities of food. Thus the efficiency of food used by the population as a whole is not merely dominated by that of the individual progeny, it is the same thing.

Now, it may be objected that agriculture is not based on mayflies and that no agricultural animals are in this category. This may not be strictly true of fish, silk moths and birds but, quite apart from this, the efficiency of *any* individual progeny may usefully indicate the theoretical maximum achievable by a population in which the costs of maintaining the parents in a reproductive state have been reduced to negligible proportions. This means that comparisons of different species (for example, cattle *vs* buffalo *vs* goats) need not be limited to the kinds of populations currently representing those animals. It becomes possible to compare the "ceiling" values (i.e. of the best individuals) above which the species cannot go without further genetic improvement of individual performance.

In the case of cattle, for example, producing less than one calf per year and living for x years, efficiency could not exceed that of the individual calf, from birth to slaughter, even if each cow produced triplets annually and lived for ever (except in the hypothetical case of progeny slaughtered very young and the dam used for meat).

Another reason for considering the efficiency of individuals is to measure and examine the very large variation that often occurs between them. This must indicate something of the potential that could be sought by genetic or nutritional improvement.

2. Family Units

The cost of producing an individual animal at birth is chiefly that of rearing and maintaining its parents, but this total cost is shared by all the progeny they produce over their lifetime. If the animal at birth is not a product, the food cost of rearing progeny to product size has to be added also.

The simplest family unit that will serve as a reasonable sample of the population is therefore one female and its progeny plus the necessary proportion of a male parent. The minimum useful period is one whole reproductive cycle (from one conception to the next) but a whole lifetime is generally better. Since males and females may not live for the same length of time, the relevant fraction of the feed used by the male has to take this into account.

In some cases, where a reproductive cycle is less than one year, an annual calculation may be satisfactory; the importance of extending the calculation to a whole-population basis depends to some extent on whether this makes a great deal of difference or not.

3. Whole Populations

The main difference is that it is necessary to take into account how the population maintains its numbers. In effect, some part of the output, that might otherwise form product, has to be devoted to replacing males and females that die. The replacement rate depends upon the mortality rate and thus on environment, disease incidence and accident. But it is usually inefficient and undesirable in agricultural systems to allow animals to die, if this can be prevented, because death contributes nothing to production. Culling, on the other hand, in the form of regular, deliberate removal of aged and inadequate animals for slaughter, can add significantly to total output: removal before animals become aged will often add even more to the output, especially in monetary terms.

Since the calculation of output may be limited to one product or may include all by-products, and all of these may be expressed in a variety of ways in relation to different or all resources, many different calculations are possible.

Even within a single kind of calculation, however, it is not possible to generalize about the relative importance of the main factors. For example, it has been suggested (Spedding, 1971) that the following formula could be used for the calculation of efficiency (E) of food conversion to carcass by whole animal populations:

$$E = \frac{(N-n)C}{D_p + D_l + D + Y(N-n) + nB}$$

where N = number of progeny reared per female
n = number of progeny required for replacement of breeding stock (per female)
C = carcass wt. of one progeny
D_p = food required by the dam during pregnancy
D_l = food required by the dam during lactation
D = food required by the dam during the non-pregnant, non-lactating period
Y = food required by each progeny
B = food required by each breeding replacement animal

This assumes no contribution to carcass output from culled animals and does not include the food requirement of the male.

A more generally useful equation would therefore be:

$$E = \frac{(N-n)P_1 + (N_c)P_2}{F_{♂} + F_{♀} + Y(N-n) + nB}$$

where P_1 = the amount of product per progeny reared
P_2 = the amount of product per female culled
N_c = the number of females culled (which must be $< n$, so that $n - N_c$ = mortality of breeding females)
$F_{♂}$ = food required by breeding males (or fraction of them)
$F_{♀}$ = total food required by breeding females

As mentioned earlier, such a calculation has to refer to a specified environment and to a sensible period of time.

It is then possible to identify those major factors that could be of great importance in determining the value of E. Clearly, increases in N can be very important, especially if n is large and when $F_♂$ and $F_♀$ are large. If the last three are small, however, N_c will be even less (because $N_c < n$), and the importance of N depends upon n.

$(N - n)$ and N_c tend to matter less if P_1 and P_2 are very large and increases in $(N - n)$ make less improvement in E as Y increases. Where either $F_♂$ or $F_♀$ are very large, improvement in E has to result from increases in $(N - n)$, N_c, P_1 or P_2.

However, it will be recognized that these factors are not independent of each other and, for example, an increase in N_c *must* be reflected in an increase in n. Similarly, P_1 can hardly be increased without also increasing Y, the food required to produce it. Other interactions are probable, between $F_♀$ and N and between P_2 and $F_♀$.

Comparisons of efficiency between animal breeds and species must take all this into account: there is no single figure that adequately represents the efficiency of a species even for one purpose.

Comparative Efficiency

For a particular purpose (and context and time period), such as the food conversion efficiency used for the previous examples, it has to be stated at what level of its potential efficiency a population is being considered.

Some examples of the relationships between E and variation in one major output factor are shown in Figs 5.9 and 5.10, for two different agricultural products. Similar relationships for natural, non-agricultural situations are shown in Fig. 5.11.

If a major common factor can be identified for several different production processes, comparisons can be made in terms of response curves relating E to changes in this factor.

This has been done in Fig. 5.12 for a number of species and products in relation to changes in reproductive rate, which appears to have a major influence in most animal production processes at the level at which they currently operate. When this is no longer the case, it generally implies that population factors are no longer of major significance: growth rate of the individual is then most likely to be the most important factor.

Most of the examples used in this chapter have naturally related to the animal species most commonly employed in agriculture. There are particular circumstances in which important contributions may be made by other species, however. In tropical Africa, for example, a whole range of rodents is eaten. These include squirrels, rats, mice, gerbils, porcupines and "grass-cutters" or cave rats (Hartog, 1973). These animals are mostly trapped, rather than farmed, but they could be the basis of more systematic and controlled forms of animal production.

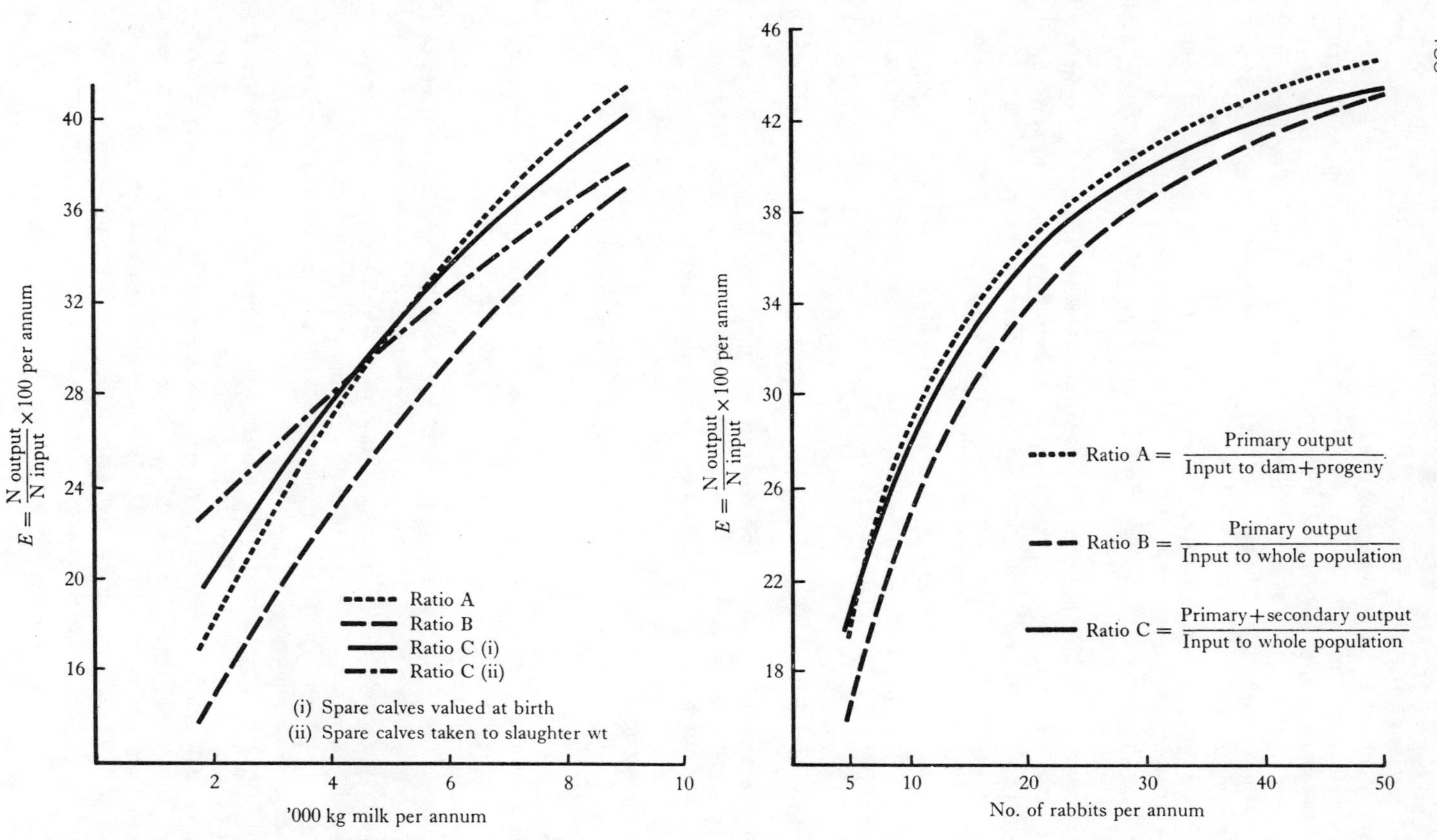

FIGURE 5.9 Population efficiency of cattle (for milk). (From Large, 1973.)

FIGURE 5.10 Population efficiency of rabbits (from Large, 1973).

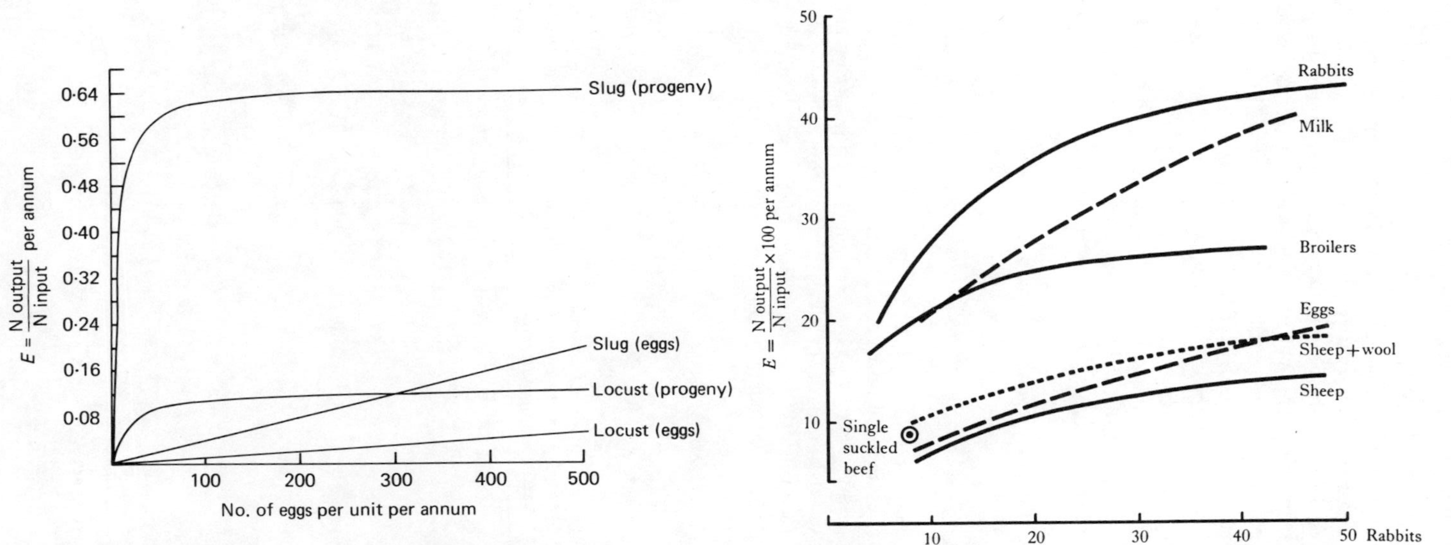

FIGURE 5.11 Efficiency of nitrogen production as body tissue from progeny and as eggs, by the locust and the slug. Curves for progeny are those for ratio

$$\frac{\text{N in progeny maturing}}{\text{Food N of parental unit + progeny}}$$

and for egg production,

$$\text{ratio } \frac{\text{Egg N}}{\text{Food N}}$$

(from Spedding, 1973).

FIGURE 5.12 Population efficiency of different species (from Large, 1973).

Since circumstances may change greatly in the future, it is as well to consider whether some of the less well known animals may be more efficient or relevant, in changed conditions, than those most used currently.

For example, the goat has often seemed a relatively neglected animal, judged to be damaging to the environment—which it is if uncontrolled (but so would other species similarly treated). It has been estimated (Teague, 1972) that goats yielding 250 gallons per head per annum could be stocked at 1·62 per acre, producing 405 gallons per acre: this compares well with cow's milk production (between 300 and 525 gallons per acre per annum).

Goats are kept for meat, milk, hair and skins and can produce on feeds of a high fibre content (Devendra and Burns, 1970): it is quite possible that they have an important contribution to make in both developed and developing countries.

There are also major possibilities for fish production.

Fish Production

Marine fishing serves as a source of food for both man and his livestock. The commercial marine fish catches include hundreds of different species but they have been dominated for many years by the Peruvian anchovy (*Engraulis ringens*). The weight of this fish caught in the period 1966-71 has varied annually between 10 and 13 million tons, some 20% of the world total. A further 25% came from the following species, each of which yielded more than 500,000 tons annually (Holt, 1973):

Alaska pollack (*Theragra chalcogramma*)
Atlantic cod (*Gadus morhua*)
Atlantic herring (*Clupea harengus*)

TABLE 5.32 Use of a trophodynamic model to estimate fish production by three ocean provinces defined according to level of primary organic production (from Regier, 1973, after Ryther, 1969)

Variable	Open ocean	Coastal zone[a]	Upwelling areas	Total
Percentage of ocean	90·0	9·9	0·1	100
Area ($km^2 \times 10^6$)	326	36	0·36	362
Mean primary production ($g\ m^{-2}$ per annum)	50	100	300	–
Total primary production ($kg \times 10^9$ per annum)	16·3	3·6	0·1	20·1
Trophic levels	5	3	1·5	–
Mean efficiency, %	10	15	20	–
Fish production ($kg \times 10^9$ fresh weight)	1·6	120	120	242

[a] Includes offshore, non-upwelling areas of high productivity.

Chub mackerel (*Scomber colias*)
Atlantic mackerel (*Scomber scomber*)
Cape hakes (*Merluccius capensis* and *M. paradoxus*)
Gulf menhaden (*Brevoortia patronus*)
Saithe (*Pollachius virens*)
Haddock (*Melanogrammus aeglefinus*)
European pilchard (*Sardinia pilchardus*)

As Regier (1973) has pointed out, fish "producers" fall naturally into two groups: aquaculturists and those who harvest organisms in the wild. For large parts of the oceans, a "hunting economy" may remain the most feasible approach: Table 5.32 illustrates the productivity of the oceans used in this way. The main disadvantage of marine fishing is likely to be the extremely high "support energy" cost of obtaining food in this way.

In many parts of the world, however, there is considerable interest in the development of aquaculture.

Fish Farming

The trend in fish farming is towards the use of several different species together. Such species have to be compatible, of course, but the main aim is that they should also be complementary, in terms of the niche they occupy or the feed component they prefer.

Examples from Indian and Chinese fish culture are shown in Table 5.33.

Expressions of productivity are greatly influenced by whether they are related to surface area or to volume of water, bearing in mind that many systems are based on flowing water. This is particularly difficult in mollusc cultures. Sessile creatures such as molluscs expend less energy on locomotion but filter

TABLE 5.33 Fish culture using complementary species (Source: Regier, 1973)

	Indian Fish Culture	
	Species	Niche
Catla	*Catla catla*	at the surface
Rohu	*Labeo rohita*	at mid-depths
Mrigal	*Cirrhina mrigala*	at the bottom
Calbasu	*Labeo calbasu*	at the bottom, feeding on molluscs
	Chinese Fish Culture	
	Species	Food
Silver carp	*Hypophthalmichthys molitrix*	plankton
Big head	*Aristichthys nobilis*	macroplankton
Grass carp	*Ctenopharyngodon idella*	coarse vegetable matter
Black carp	*Mylopharyngodon piceus*	molluscs
Mud carp	*Cirrhina molitorella*	worms and organic mud
Common carp	*Cyprinus carpio*	an omnivorous scavenger

their food out of substantial volumes of water, often within highly fertile estuaries.

Rafts of mussels (*Mytilus edulis*) in some Spanish estuaries, for instance, have been reported (Pinchot, 1970) to produce annually about 300 000 kg of edible flesh per hectare (of raft).

In addition to the harvesting of such controlled populations living on the naturally occurring food supply, several attempts have been made to improve production by enrichment with chemical fertilizers. This has been successful in freshwater ponds but less clearly so in sea-water inlets, such as the Scottish sea lochs (Weatherley, 1972). An example of successful enrichment may be found in the studies of a Tasmanian trout lake (Weatherley and Nicholls, 1955).

Fish farming in freshwater creates rather more problems in terms of water supply but both fresh- and salt-water farming are much influenced by the length of the growing season, determined mainly by temperature. This makes the tropics and subtropics extremely suitable for fish production of all kinds.

Carnivorous fish, such as trout and eels, are farmed, being fed with slaughterhouse offals and trash fish from the fish markets (Hickling, 1966).

Omnivores such as the carp family are farmed intensively on food produced naturally, with or without fertilizer, or on added fodders, including oilcakes and maize residues. Productivity may be very high and Hickling (1966) quotes 800 lb per acre per annum in Germany and Russia and over 4000 lb in the Far East. Israel has recorded (in 1964) a country average of 1860 lb of fish per acre per annum. When concentrates are used, however, fish and poultry production are in direct competition and it may be better to use fertilizer.

Phosphates are the main fertilizers for fish production and, in the tropics, a response of 15 lb of fish has been obtained for each lb of P_2O_5. With phosphates alone, fish ponds in Malacca have produced between $\frac{3}{4}$ and 1 ton of fish per acre per annum.

Marine fish may also be cultivated, as with the yellowtail in Japan (Loftas, 1969). Young fish are caught in the open sea and transferred to enclosed areas of the sheltered inland sea for "fattening". In recent years there have been many developments in the rearing of young fish in tanks, for later transfer to enclosed sea lochs, but it is not only fish that have been the subject of marine farming.

Marine Farming of other Animals

Sea-water ponds are used in Japan for cultivating the prawn (*Penaeus japonica*) and the cultivation of oysters (*Ostrea edulis*) is both widespread and long-established. In 1967, the total harvest of oysters was 0·83 m tons, of squids 0·75 m tons, of shrimps and prawns 0·69 m tons, and clams and cockles produced 0·48 m tons (Holt, 1969).

There have been recent developments towards farming operations with turtles and it may be expected that attempts will be made to extend cultivation practices to other animals hitherto only hunted.

Turtle farming is an example of the attempt to combine control of critical phases, such as early rearing, with exploitation of large water masses by animals with definite reproductive patterns associated with specific locations. Thus, although the animals derive their food from a very large area, it is known where they will breed and their eggs can be collected. Incidentally, their eggs are reckoned to be very high in protein (up to 29% on a fresh weight basis) and thus of considerable nutritional value (Somadikarta and Anggorodi, 1962).

The collection of turtle eggs has been especially prevalent in the Muslim areas of Malaysia whereas, in other parts of the world, harvesting has been mainly of the adults for flesh and fat (Hendrickson, 1958). Hendrickson (1958) pointed out that the harvesting of eggs constitutes exploitation at that stage of the life cycle where the species is adapted to sustain high losses in the natural course of events. In this sense, many of the eggs collected represent a diversion of inevitable losses. Since average female Sarawak green turtles (*Chelonia mydas*) lay some 600 eggs per season for at least three seasons and only about 0·1% are needed to maintain the population, the potential is considerable.

Under farming conditions, of course, the situation is different and the production of turtle meat, based on the herbivorous green turtle, appears promising (Carr, 1952; Bustard, 1972). Growth rates can be much higher under farming conditions than in the wild, turtles achieving 27-36 kg weight at 2 years of age and 54 kg by the time they are 3 years old: this is from an egg weighing 36-52 g.

Another salt-water animal beginning to be farmed is the crocodile (*Crocodylus porosus*) and growth rates of this species are also reported to be higher under farming conditions (Bustard, 1972).

References

ARC (1963). Nutrient requirements of Farm Livestock. No. 1. Poultry.

ARC (1965). Nutrient requirements of Farm Livestock. No. 2. Ruminants.

ARC (1966). Nutrient requirements of Farm Livestock. No. 3. Pigs.

Asdell, S. A. (1964). "Patterns of Mammalian Reproduction" (2nd edition). Constable, London.

Baker, R. D. (1973). Personal communication.

Bannerman, M. M. and Blaxter, K. L. (1969). "The Husbandry of Red Deer". Rowett Res. Inst. and Highlands and Islands Dev. Board.

Barker, J. M. (1973). Personal Communication.

Ben Shaul, D. M. (1962). International Zoo. Yearbook **4**, 333-342.

Blake, C. D. (ed.) (1967). "Fundamentals of Modern Agriculture". Sydney University Press.

Blaxter, K. L. (1962). "The Energy Metabolism of Ruminants". Hutchinson, London.

Bonadonna, T. (1966). *Wld Rev. Anim. Prod.* **3**, 46-68.

Borgstrom, G. (1969). "Too Many". Collier-Macmillan, New York.

Boyazoglu, J. G. (1963). *Ann. Zootech.* **12** (4), 237-296.

Brody, S. (1945). "Bioenergetics and Growth". Reinhold, New York.

Bustard, R. (1972). "Sea Turtles". Collins, London and Sydney.
Campbell, A. G. (1968). *Span* **11** (1), 50-53.
Carr, A. (1952). "Handbook of Turtles". Comstock, New York.
Castle, M. E. and Holmes, W. (1960). *J. agric. Sci.* **55** (2), 251-260.
Charles, D. D. and Johnson, E. R. (1972). *Aust. J. agric. Res.* **23**, 905-911.
Cockrill, W. Ross (1967). *Wld Rev. Anim. Prod.* **3** (13), 98-107.
Cockrill, W. Ross (1969). *Farmer's Weekly.* January 3rd.
Colburn, M. W. and Evans, J. L. (1968). *J. Dairy Sci.* **51**, 1073-1076.
Commonwealth Secretariat (1970). "Industrial Fibres. A Review".
Conway, A. (1968). *Span* **11** (1), 47-49.
Crawford, M. A. (1968). *Vet. Rec.* **82** (11), 305-314.
Crile, G. and Quiring, D. P. (1940). *Ohio J. Sci.* **XL** (5), 219-259.
Cross, B. A. and Harris, G. W. (1952). *J. Endocr.* **8**, 148.
Devendra, C. and Burns, M. (1970). "Goat Production in the Tropics". Commonwealth Agricultural Bureaux, Farnham Royal, Bucks.
Dickerson, G. (1970). *J. Anim. Sci.* **30** (60), 849-859.
Dodd, C. (1972). *The Guardian.* June 26th.
Donald, H. P. and Russell, W. S. (1970). *Anim. Prod.* **12** (2), 273.
Duckham, A. N. and Lloyd, D. H. (1966). *Farm Economist* **11** (2), 95-97.
El Naggar, A. A., El Shazley, K. and Ahmed, I. A. (1970). *Anim. Prod.* **14** (2), 171-176.
Epstein, H. (1969). "Domestic Animals of China". Commonwealth Agricultural Bureaux, Farnham Royal, Bucks.
Ewbank, R. (1967). *Anim. Behav.* **15**, 251-258.
Ewbank, R. (1969). *Br. vet. J.* **125** (2), IX-X.
FAO (1970a). Commodity Bulletin, Series 48. The World Hides, Skins, Leather and Footwear Economy. FAO Rome.
FAO (1970b). *Nutr. Stud.* No. 24.
FAO (1971a). *Production Yearbook* 1970, **24**.
FAO (1971b). Rep. of an ad hoc consultation on the value of non-protein nitrogen for ruminants consuming poor herbages. Kampala, Uganda, 29 June-3 July 1971.
FAO (1972). *Production Yearbook* 1971, **25**.
Fishwick, V. C. (1953). "Pig Farming". Crosby Lockwood, London.
Fraser, A. F. (1971). "Animal Reproduction; Tabulated Data". Bailliere, Tindall and Cox, London.
French, M. H. (1966). *Wld Rev. Anim. Prod.* **3**, 89-94.
French, M. H. (1970). "Observations on the Goat". FAO Agric. Studies. No. 80, p. 189.
Hafez, E. S. E. (1962). "The Behaviour of Domestic Animals". Bailliere, Tindall and Cox, London.
Hall, G. O. and Marble, D. R. (1931). *Poultry Sci.* **10**, 194.
Hartog, A. P. den (1973). FAO Nutrition Newsletter **11** (2), 1-14.
Hawthorne, G. R. (1972). Personal communication.
Hendrickson, J. R. (1958). *Proc. zool. Soc. Lond.* **130**, 456-534.
Hickling, C. F. (1966). *Proc. Nutr. Soc.* **25** (2), 140-144.

Hickling, C. F. (1971). "Fish Culture". (2nd edition) Faber and Faber, London.
Hintz, H. F., Schryver, H. F. and Halbert, M. (1973). *Anim. Prod.* **16** (3), 303-305.
Holmes, W. (1970). *Proc. Nutr. Soc.* **29** (2) 237-243.
Holmes, W. (1971). *In* "Potential Crop Production". (P. F. Waring and J. P. Cooper, eds). Heinemann, London.
Holt, S. J. (1969). *Scient. Am.* **221** (3), 178-194.
Holt, S. J. (1973). Symp. "The Man/Food Equation", Sept. 1973, Lond.
Hope, H. (1972). *Farmer's Weekly*. **August 11th (IV) and (V).**
I.D.F. document. IX DOC. 5 59/28.
Jenness, R. (1972). *Hannah Res. Inst. Rep.,* **1972, 47-55.**
Johnstone-Wallace, D. B. and Kennedy, K. (1964). *J. agric. Sci. Camb.* **34**, 190.
Kenneth, J. H. and Ritchie, G. R. (1953). "Gestation Periods" (3rd edition). Commonwealth Agricultural Bureaux, Farnham Royal, Bucks.
Kleiber, M. (1947). *Physiol. Rev.* **27**, 511.
Kosin, I. L. (1962). *In* "Reproduction in Farm Animals" (E. S. E. Hafez, ed.) Bailliere, Tindall and Cox, London.
Krutyporoh, F. and Treus, V. (1969). *Moloch. myas Skotov, Mosk.* **14** (8), 36-38 (in Russian).
Large, R. V. (1973). *In* "The Biological Efficiency of Protein Production". (J. G. W. Jones, ed.) Cambridge University Press.
Large, R. V. (1974). Personal communication.
Lawrie, R. A. (ed.) (1970). "Proteins as Human Food", p. 144. Butterworths, London.
Lawton, J. H. (1973). *In* "Resources and Population". (B. Benjamin, P. R. Cox and J. Peel, eds) pp. 59-76. Academic Press, London and New York.
Ledger, H. P., Sachs, R. and Smith, N. S. (1967). *Wld. Rev. Anim. Prod.* III (11), 13-37.
Leupold, T. (1968). *Anim. Breed. Abstr.* (1970). **38** (4), p. 669.
Linzell, J. L. (1972). *Dairy Sci. Abstr.* **34** (5), 351-360.
Loftas, T. (1969). "The Last Resource". Hamish Hamilton, London.
Mackenzie, D. (1972). "Goat husbandry" (3rd edition). Faber and Faber, London.
Mason, I. L. (1951). "A World Dictionary of Breeds, Types and Varieties of Livestock". Tech. Comm. No. 8. Commonwealth Agricultural Bureaux, Edinburgh.
Mason, I. L. (1967). "The Sheep Breeds of the Mediterranean". Commonwealth Agricultural Bureaux, Farnham Royal, Bucks.
McArdle, A. A. (1966). "Poultry Management and Production" (2nd edition). Angus and Robertson, Sydney.
McBee, R. H. (1971). Ann. Rev. Ecol. and Systematics, **2**, 165-176.
Morris, T. R. (1971). Personal communication.
Nickel, R., Schummer, A. and Seiferte, E. (1973). "Viscera of the Domestic Mammals". (Translated and revised by W. O. Sack). Verlag Paul Parey, Berlin and Hamburg.

Olsson, N. O. (1969). *In* "Nutrition of Animals of Agricultural Importance". (Sir D. Cuthbertson, ed.), Part 2, Chapter 25. Pergamon Press, Oxford.

Ontario Dept. of Agric. "Duck and Goose Raising". Publ. 532.

Orraca-Tetteh, R. (1963). *Inst. Biol. Symp. No. 10*, 53-61.

Paquay, R., de Baere, R. and Lousse, A. (1972). *Br. J. Nutr.* **27**, 27-37.

Parker, A. (1947). *New Biology* No. 3, 149-170.

Petrides, G. A. and Swank, W. G. (1965). *Proc. 9th Int. Grassld Cong. Brazil,* **1**, 831-841.

Pinchot, G. B. (1970). *Scient. Am.* **223**, 15-21.

Pirie, N. W. (1971). "Leaf Protein". I.B.P. Handbook No. 20, Blackwells, Oxford.

Porter, W. P. and Gates, D. M. (1969). *Ecol. Mon.* **39**, 245-270.

Pynaert, L. (1951). "Le Manioc" (2nd edition). La direction de l'agriculture, Minisere de Colonies, Belgique.

Raistrick, A. S. (1973). Personal communication.

Regier, H. A. (1973). *In* "The Biological Efficiency of Protein Production" (J. G. W. Jones, ed.), pp. 263-279. Cambridge University Press.

Robb, J., Harper, R. B., Hintz, H. F., Reid, J. T., Lowe, J. E., Schryver, H. F. and Rhee, M. S. S. (1972). *Anim. Prod.* **14** (1), 25-34.

Rolfe, E. J. and Spicer, A. (1973). *In* "The Biological Efficiency of Protein Production" (J. G. W. Jones, ed.), pp. 363-370. Cambridge University Press.

Roy, J. H. B. (1959). "The Calf". Vol. I. Iliffe.

Ruthenburg, H. (1971). "Farming Systems in the Tropics". Oxford University Press.

Ryther, J. H. (1969). *Science* **166**, 72-76.

Schulthess, von W. (1967). *Wld Rev. Anim. Prod.* **3** (13), 88-97.

Somadikarta, S. and Anggorodi, R. (1962). *Communicationes veterinarial* **6**, 73.

Spedding, C. R. W. (1970). "Sheep Production and Grazing Management" (2nd edition). Bailliere, Tindall and Cassell, London.

Spedding, C. R. W. (1971). "Grassland Ecology". Oxford University Press.

Spedding, C. R. W. (1973). *In* "The Biological Efficiency of Protein Production". (J. G. W. Jones, ed.) Cambridge University Press.

Spedding, C. R. W. and Hoxey, A. M. (1974). Proc. 21st Easter School, Nottingham. Butterworths, London.

Takeuchi, Y. (1967). *In* "Introduction to Silkworm Rearing", Part III, p. 63. The Japan Silk Ass. Inc., Tokyo.

Talbot, L. M., Payne, W. J. A., Ledger, H. P., Verdcourt, L. D. and Talbot, M. H. (1965). "The Meat Production Potential of Wild Animals in Africa". Tech. Commun., No. 16, Commonwealth Agricultural Bureaux, Farnham Royal, Bucks.

Taylor, R. J. (1969). "Facts about Margarine", pp. 28-29. Van den Berghs Ltd.

Taylor, St. C. S. (1965). *Anim. Prod.* **7** (2), 203-220.

Teague, P. J. (1972). B.Sc. Hons Dissertation, University of Reading.

Treacher, T. T. (1969). *Grassland Research Institute Int. Rep.* **159**.

Turner, H. N. (1971). *Outlook on Agriculture,* **6** (6), 254-260.

U.S. Dept. of Int. (1963). "Fish and Wildlife Service. Trout Feeds and Feeding". Circular 159.

Vaccaro, R., Dillard, E. U. and Lozano, J. (1968). A.L.P.A. Mem. 3. 115-126.

Vanschoebroek, F. and Cloet, G. (1968). *Wld Rev. Anim. Prod.* **IV** (16), 70-76.

Walsingham, Jean M. (1972). Unpublished data.

Walsingham, Jean M. (1973). Personal communication.

Walsingham, Jean M. (1974). Personal communication.

Wassmuth, R. (1972). "The Biological Efficiency of Meat Production in Sheep". E.A.A.P. Sheep and Goat Commission. Verona.

Waterman, J. J. (1968). "The Cod". Torry Advisory Note, No. 33.

Weatherley, A. H. and Nicholls, A. G. (1955). *Aust. J. mar. Freshwat. Res.* **6**, 443-468.

Weatherley, A. H. (1972). "Growth and Ecology of Fish Populations". Academic Press, New York and London.

Wilson, P. N. (1968). *Chemy Ind.* 899-902.

Yokoyama, T. (1963). *A. Rev. Ent.* **8**, 287-306.

Young, E. and Van Den Heever, L. W. (1968). *Jl S. Afr. vet. med. Ass.* **40** (1), 83-88.

6 Harvesting the Product

The word "product" may be used for a number of different stages leading up to the *final* product. "Harvesting" is usually applied to the plant and animal populations from which products are derived but it is not generally the final products that are harvested.

This is easily visualized and easily described for plant production. Here, plant growth results in plant populations that at some time include material regarded as "the crop". Harvesting and cropping are then synonymous. The crop may well require further processing before the final product is arrived at, but the product of crop production is the crop. In animal production there is no precisely equivalent term, perhaps because whole animals are often harvested from a population. The animal and the product harvested are then the same, although not representing the final product. On a population basis there is more similarity and sheep farmers commonly refer to the "lamb crop" in the same way as a fruit farmer refers to the "apple crop", both long before harvesting.

In milk, egg and wool production, harvesting is comparable to that of plants and, significantly, the word crop is often applied to wool (the wool crop referring to the wool produced by a sheep population, rather than to that of an individual).

The word most commonly used for both plants and animals is "yield" and terms like "milk yield" and "bean yield" are unambiguous. There is some ambiguity, however, in the use of yield for output of product per head and per unit of land. Wool yield could be applied to the fleece weight of one sheep but yield of crops is usually applied to areas of land. This is not entirely so, however, especially with fruit trees and glasshouse crops, and, strictly speaking, yield simply means "quantity yielded" and must be qualified in relation to a resource used. Thus yield could be per hectare, per herd, per unit of irrigation water or nitrogenous fertilizer, per cow or per man.

Once the yield, crop or product has been harvested, it may be subjected to all kinds of processes before it is sold and more still before it is consumed. Wool may be washed before selling, and dyed, spun and woven before being used. Meat has to be separated from the inedible parts of the whole animal, though not necessarily before sale, and cooking takes place before consumption. Sugar beet is extracted, cabbages have outer leaves removed and are later cooked, fruit is ripened and may be canned, and potatoes may be sliced and cooked before or after sale.

There has been some tendency to combine operations: the combine harvester collects the cereal crop, threshes it, sacks the grain and rejects the straw in one operation; pea harvesters now take away peas and leave haulm and pods behind.

In this account of agricultural biology most of these processes will be ignored, but there are two points to be noted. First, processing is of increasing importance and has a large effect on the quantity and quality of agricultural produce. Milk, meat, eggs and fish may be used fresh, heat-treated or cooked, processed as cheese, yoghurt or dried milk, or combined with other foods (Aylward and Hudson, 1973). Vegetables and cereals, similarly, may be used fresh or cooked, processed by canning, freezing or dehydrating, or mixed with other foods.

Secondly, some processing is essentially biological. The conversion of milk into cheese by the use of bacteria is an obvious example; hanging of meat and fermentation of barley may be less so.

Harvesting will be considered here, then, as the initial process of collecting the first products from the plant and animal populations kept to produce them, and the immediate processing that is used to produce a saleable product from the farm.

Viewed in this way, there are four main harvesting categories:

(1) direct harvesting of primary production;
(2) indirect harvesting of primary production;
(3) direct harvesting of secondary production; and
(4) indirect harvesting of secondary production.

The last one gives rise to a fifth category, direct harvesting of tertiary production, but it is unimportant and best dealt with under (3) and (4).

In each case, harvesting may or may not affect the production process and in each case will result in some loss and wastage. In general, it is the object of harvesting to minimize losses and wastage and to optimize the proportion harvested at any one time in relation to the effect on subsequent production. All of this is done, however, within an economic framework, and the output of a harvesting method has to be related to its cost and effort. It is usually not possible, therefore, to *minimize* losses, except within certain constraints, and some production processes are based on the use of wastage from others.

1. Direct Harvesting of Primary Production

Where a plant population only produces one crop at one time, the main object of harvesting methods must be to collect as high a proportion of the product as is consistent with the cost of collection. This applies to grain harvests, for example; clearly the collection of every grain would be uneconomic but the more that is gathered, without incurring disproportionate costs, the better.

However, even here there are other considerations if the system examined is enlarged a little. Since the plant population is not going to be involved again in a production process it does not matter if it is damaged. But it may matter whether or not the soil is damaged, or whether a lot of surface litter is left, or whether weed seeds are scattered, or whether crop seeds are left that will interfere with the growth of the next crop.

So harvesting methods may have other effects within the larger system of land use. Where a plant population produces successive crops, however, the situation may be quite different. Consider an apple tree, for instance: clearly the effects of harvesting method on the well-being of the tree and, in particular, on its longevity and capacity to produce subsequently, are of great importance.

A different situation occurs where the crop product is not a specialized part of the plant, in the sense that seeds and tubers are specialized organs, often concerned with storage.

Where the crop is a foliar one, for example, as with grass or tea, and can be harvested at intervals over many years, both longevity and productivity in the short term can be markedly affected by harvesting methods. Furthermore, it is not merely the *methods* that matter but also the amount harvested.

In plants where all the leaves are suitable for cropping, the total removal of all of them at any one time represents the height of harvesting efficiency; but it may result in reduced productivity, or even death of the plant. Thus, forage grasses and legumes grow less well and yield less dry matter, energy and digestible organic matter annually if harvesting is too severe (see Spedding and Diekmahns, 1972). Severity, in this context, generally implies both frequent harvesting and the removal of a high proportion of the leaf on each occasion.

If all the leaf (but not, of course, all the plant) is removed at harvest, harvests must be infrequent: if harvests occur at short intervals, less must be removed at each one.

Curiously enough, there are cases of an almost opposite kind. For instance, plants of wheat (*Triticum aestivum*) or Italian ryegrass, if allowed to grow uninterruptedly, are annuals, but if they are defoliated repeatedly, so that they do not flower, they behave as perennials, remaining in a short, leafy, vegetative state.

Perennial ryegrass (*Lolium perenne*) is normally long-lived but may fail to survive low winter temperatures, especially if it enters the winter with a great deal of leaf. Survival is improved if such swards are defoliated in the autumn.

In all these cases there is an optimum harvesting pattern involving method, equipment, frequency and severity, for maximizing harvested annual (or lifetime) yield. These patterns will vary between crops and climates and may not be constant within a year. For example, it may be best to relate harvesting frequency to the rate of plant growth or leaf production, in which case it will have to vary with temperature, water supply, fertilizer usage and light incidence. If the optimum pattern is also defined in terms of the amount or proportion removed at any one time, and if these amounts or proportions also vary with growth stage and season, it is clearly no easy matter to define the optimum for all purposes.

There is often a compromise in practice, therefore, between the biological optimum and what is practicable: it is no use in practical agriculture trying to relate practice to criteria that cannot be measured easily on the farm.

2. Indirect Harvesting of Primary Production

This occurs where animals are used to harvest the crop and, generally, to convert it to animal products.

There are just a few examples, incidentally, where animals may be said to be employed *only* as collectors. It is perhaps a marginal one in the case of the honeybee (*Apis mellifera*): bees certainly collect nectar, and its conversion to honey hardly involves the metabolism of the bee, being a straightforward conversion to simpler sugars in the bee's "honey-stomach" (see Butler, 1954).

Recent studies of the energetics of this kind of "collecting" activity have been carried out on the bumblebee *Bombus ternarius*. It seems that a bee may have to visit more than 200 flowers to collect about 100 μl of nectar. Bumblebees work from dawn to dark and can operate at 0°C by exercising the flight muscles before take-off, in order to generate heat (Heinrich, 1973). These bees are primarily collectors of pollen, however, and their agricultural importance is as pollinators of agricultural crops (Free and Butler, 1968).

The crux of this question is, of course, how to use an animal purely for collecting, when it will generally only collect what it wishes to eat. There are, therefore, only three possibilities. One is to train animals to collect what they would otherwise not be interested in: this may also meaning training to catch and kill as well as collect. A second possibility is to recover what they have collected and eaten: the regurgitation of whole fish by birds is an example (though not relevant to harvesting of primary production) and more bizarre ones can be found where birds are killed and their crop contents removed for cooking and consumption. An isolated example of the latter is referred to by Gilbert White (Johnson, 1931) but this was incidental to killing a pigeon for its flesh. Seed-eating birds are sometimes used in this fashion, however.

Thirdly, there are animals that collect food for storage. Squirrels do this but I do not know of any examples of the habit being systematically exploited.

For the most part, however, indirect harvesting by animals results in animal products, and the crop that is harvested is that part of the plant population that animals use for food. It is possible to persuade animals to harvest material that is not a complete food for them, by supplementing the diet with those constituents in which it is deficient. Pastures of very poor quality, on the South African veldt, for example, can be used by cattle given a supplement of urea as an additional source of nitrogen (FAO, 1971; Lampila, 1972). This is an issue of some importance, because there is much less point in using animals to harvest what we could eat and harvest ourselves.

Thus pigs may be fed on barley because we want the meat but they are not used to harvest barley; they *are* (or were) used to harvest acorns and beechmast, however, partly because we do not wish to use these seeds directly and partly because of the difficulties and cost of direct harvesting.

When plants produce storage organs, such as seeds, roots and tubers, it becomes much easier to harvest the crop directly. This is partly because the products are more concentrated and less bulky and are therefore easier to locate, collect, handle and transport. For similar reasons, they are also more likely to be suitable for human consumption.

There are many reasons, therefore, why the animals most commonly used for indirect harvesting are grazers and browsers that live primarily on plant leaves. It may be noted here that man as a leaf-eater tends to confine himself to juvenile leaves, preferably concentrated in buds (e.g. cabbage), unless a further extraction process is intended (as with tea).

It is perhaps surprising that so little agricultural use has been made of tree-browsing animals, for it might be imagined that leaf production by trees could be highly productive and that browsing animals would be much more efficient as harvesters than any mechanical device. As it is, the majority of agricultural leaf-eaters are grazers. Similar arguments apply here, in relation to defoliation frequency and severity, as were noted in the previous section, but there are, in addition, many other factors associated with the presence of the animal. These are the effects of treading, lying and fouling with faeces and urine. Some of these effects are straightforward losses of material already grown and otherwise harvestable and these are discussed further in the next chapter. Other effects are indirect losses of production due to a reduction in growth rate, at the time or subsequently.

In fact, of course, any of these factors could cause losses of both kinds. If faeces or urine are present in large enough amounts, they are bound to influence the quantity available for harvesting, the rate at which it is lost and the rate of herbage growth. They do not always operate in all these ways, however. In a particular system at a particular time, the effect of urine may be quite transitory, the effect of faeces may be to reduce the amount of harvestable herbage and the effect of treading may be entirely on subsequent regrowth. In the same grazing system, at a different time or in different weather conditions, the situation might be quite different in all these respects.

A major consequence of the method of indirect harvesting of primary production is the effect on the productivity of the harvesting animals themselves. The performance of the latter depends initially on their genetic capacities and on the ability of the environment to supply their needs. Thus performance will generally reflect the quantity of nutrients ingested in relation to what is required, and feed intake will be influenced by both the quality and the quantity of feed available. Some care is needed in the interpretation of what is "available" because the amount eaten and the energetic cost of obtaining it are affected by how hard the animal has to work in order to harvest it.

Management of pastures (and this includes the imposition of methods of grazing) can greatly influence herbage quality, and this affects both the intake of herbage by animals and the nutritive value of what they consume. In general terms, there exists an important relationship between feed intake and digestibility (Blaxter, 1962), where digestibility is being used as the most useful single expression of nutritive value (Fig. 6.1). However digestible or desirable the herbage is, intake is bound to be related in some fashion to the amount present. For herbivorous animals the food is relatively static in space but dynamic with respect to time.

Availability cannot be described without using gradations and, as mentioned

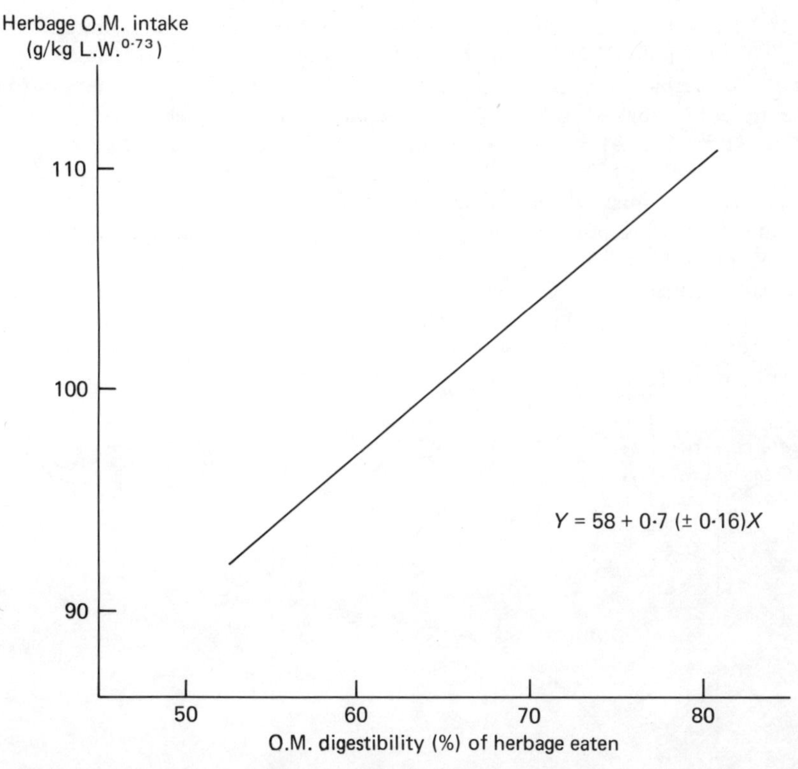

FIGURE 6.1 The relationship between herbage intake and the digestibility of the herbage selected; data for 5–9 month-old Friesian calves grazing the primary growth of S24 perennial ryegrass (Rodriguez and Hodgson, 1974).

earlier, this leads to many difficulties of interpretation. Nevertheless, there is often a fairly clear distinction between what is *unavailable*, i.e. no effort on the part of the animal allows access to it (and for this reason commonly described as inaccessible), and what can be consumed by normal activity. Of course, we may then consider what is "normal" and there are certain to be boundary cases where we cannot be sure whether food is available or not.

Where the distinction is reasonably clear, food intake (I) must be determined by the time (t) spent harvesting, the amount ingested per bite (b) and the number of bites (n) per unit of time:

$$I = b \times n \times t$$

Since I will have an upper value representing the maximum voluntary intake of the animal on that food and at that time (t), a ratio can be used to indicate the extent to which the maximum is being achieved:

$$\frac{I_t}{I_{\max}}$$

If both animal and non-herbage environment remain constant, variations in this ratio (or indeed in I) must reflect changes in b, n or t. If the quantity and quality of herbage remain constant, variations in food intake would seem to be due to either environmental or animal changes. This situation can be readily imagined for, say, a leaf-eating caterpillar. Unfortunately, it can also be imagined for sheep and cattle and yet does not appear to be true. That is to say that "quantity and "quality" of herbage proves to be an inadequate description of a pasture from this point of view. Thus the spatial distribution of foliage is important to the grazing animal as well as to the growing plant.

Spatial distribution is influenced by plant form, of course, and the structure and growth habit of the plant greatly influence the selectivity of grazing by animals.

One major consequence of all this is that the food intake (I) of grazing cattle

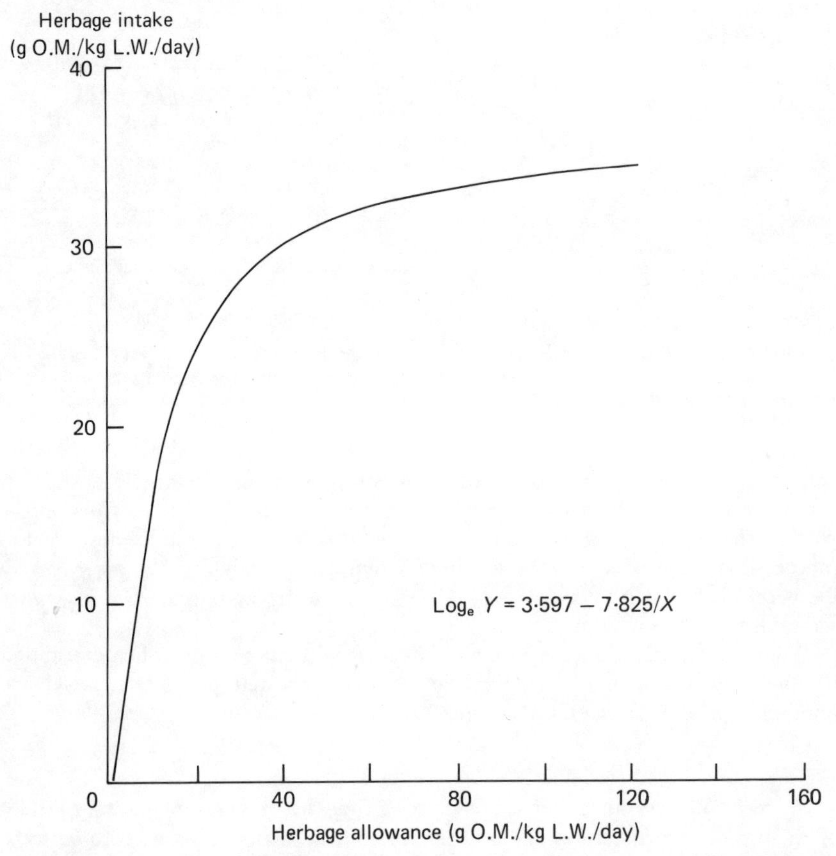

FIGURE 6.2 The relationship between herbage allowance (S23 perennial ryegrass) and herbage intake in artificially-reared lambs (3-6 months old) (from Gibb and Treacher (1974)).

and sheep varies with the quantity of herbage present on each unit area of land (see Fig. 6.2). (The need to convey a relatively simple picture of this situation should not obscure such difficulties as defining the meaning of "what is present" in relation to time.) It is not possible, therefore, to divide the amount of herbage available by the amount required by each animal and thus arrive at the optimum number of animals to employ in harvesting, even over a short period.

Indeed, it is often not possible both to harvest all that is available and, at the same time, ensure that each animal obtains what it requires (another useful ratio is I/R, where R = food required).

This leads to the major dilemma in grazing management:

(a) Stated in research terms—how to combine maximum food intake per animal and maximum utilization of herbage;
(b) Stated in practical terms—what is the optimum compromise between food intake per animal and harvesting efficiency?

An example of the problem is shown in Fig. 6.3, where animal performance per head and per hectare move in opposite directions when related to the number of animals per hectare.

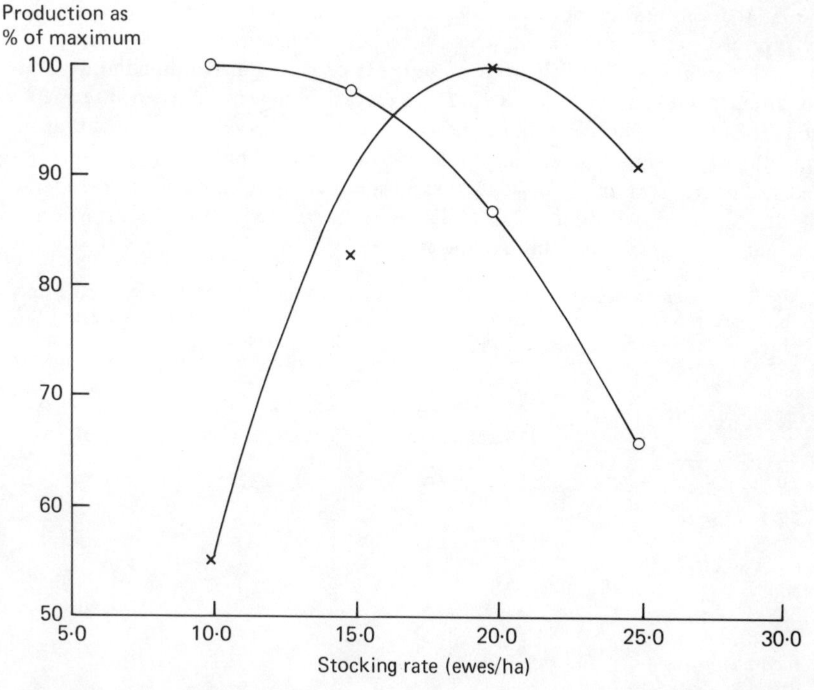

FIGURE 6.3 An example of the relationship between production per hectare and growth rate of the individual animal (derived from Spedding (1970)). X = mean l.w. gain of lambs per hectare against stocking rate of ewes; O = mean l.w. gain per lamb per day against stocking rate of ewes.

It is not only grazing animals that show this tendency, however. Animals fed indoors will frequently consume less than their maximum voluntary food intake per day if they are offered no more than that (I_{max}) (Blaxter *et al.*, 1961). To be sure that a sheep will consume 100% of its maximum voluntary intake, it is necessary to offer it approximately 115%.

One interesting difference between indoor and outdoor situations is that this necessary surplus can be provided by living herbage in the latter case. It is no great disadvantage, therefore, that it is not eaten, since it will be there tomorrow and even continue to grow. Indoors, there may be unavoidable wastage.

3. Direct Harvesting of Secondary Production

By no means all of the secondary production represented by herbivorous animals is agriculturally useful. It is true that a great deal of every animal is used, one way or another, but the main product may be less than half of what has been grown.

For meat production, one of the commonest expressions of usefulness of the live animal is the "killing-out percentage" (K.O. %) or "dressing-out percentage" (D.O. %). These terms are synonymous and refer to the proportion of the live animal that occurs in the dressed carcass after slaughter (see Table 7.6). There are several ways of formulating the expression, based on cold or warm carcasses and the condition (e.g. fully fed or starved) of the animal immediately before slaughter. The essential point is that the carcass represents a major intermediate product between the live animal and edible meat. It may seem curious that such a major product ignores several edible and other useful parts of the animal body (e.g. the heart, liver and skin of cattle and sheep) and yet includes a considerable amount of inedible bone. Table 6.1 illustrates the relative proportions of different tissues in the live animal and its carcass.

TABLE 6.1 The weight of body components of animals as a percentage of their empty body weight (E.B.W.) (Prepared by T. T. Treacher from Pálsson and Verges, 1952; Treacher and Hodgson, 1970; Moulton *et al.*, 1922; McMeekan, 1940)

Animal	Lamb		Ewe	Calf	Steer		Pig
Age	9 weeks	41 weeks	Mature	16 weeks	11 weeks	21 weeks	16 weeks
E.B.W. (kg)	24	76	60	68	232	324	47
% of E.B.W. as:							
Carcass	59	65	61	57	70	75	69
Head and feet	7	5	6	26			7
Hide	11[b]	10[b]	8[c]		10	9	2
Internal fat	2	8	10	0			1
Edible organs	14	8	3	3	14	10	4
Inedible organs			8	13			11
Blood	7	4	4	a	6	6	6

[a] This value is included in head, hide etc.
[b] Including wool
[c] Shorn

TABLE 6.2 Proportions harvested and consumed: selected illustrations for a few products

Product	Amount harvested	Yield of product	Amount edible	Amount consumed	Source
Beef	432 kg live weight	250 kg carcass plus edible offal 58 kg	184 kg carcass less bone + 58 kg offal	140 kg carcass less bone + trimmed fat; 58 kg offal. Cooking losses —water and fat.	Gerrard (1971)
Rabbit	2 kg live weight	1 kg carcass	794 g raw meat	567 g cooked meat	MAFF (1971)
Milk	3864–11365 litres per cow per year →	→	→	Losses include film of milk left on each piece of equipment in production and processing. Vitamins C and B due to exposure to light, and storage.	
Honey	11–27 kg + per hive per year →	→	→	Processing losses on equipment, etc.	
Fish (trout)	200 g live weight	180 g carcass	80 g edible flesh	80 g less cooking losses, probably fat and water.	USDI (1963)
Egg (hen)	58 g whole egg	52 g (shell removed)	52 g	Losses in cooking vary from 0 for boiled egg to 8–13% for fried or scrambled.	MAFF (1971)

The carcass is, of course, an object of trade and is generally further subdivided, into joints, before sale to the consumer. A considerable amount of "wastage" occurs at the time of consumption and in the food preparation stage and, in addition, a further necessary separation of edible from inedible parts.

It has been estimated that, of the original live animal, perhaps only 30-40% of the energy content is actually eaten as human food. Estimates of the efficiency of animal production should take this into account at some stage, but similar considerations apply to crop production (for example, how much of the harvested carrot or potato do we actually eat?). From this point of view, eggs and milk are quite remarkable (see Table 6.2) in the high proportion of the harvested product that is edible. It is a rather different matter, however, as to whether *harvesting* is efficient.

In the case of milk, the amount produced and the amount harvested are necessarily the same, so the question is really whether more could be produced if harvesting methods were changed. This is usually so: more frequent milking, for example, would normally allow higher milk production. It is worth noting that better or more feed will not result in more milk if the rate of milk removal is, in fact, the limiting factor. Of course, as with many other harvesting procedures, it does not follow that increased harvesting frequency would be more profitable, because the costs remain high and the additional production may be small. With eggs it is different and methods of collection rarely influence production. They have their effect on wastage, however, by modifying loss by breakage.

Harvesting of wool involves shearing in the case of most sheep, though a few breeds are plucked, such as the Improved Shetland sheep on which the Fair Isle woollen industry is based: Llama, Alpaca, Vicuna, Huanaco, Angora rabbits and goats are also plucked. Table 6.3 shows the range of wool yield of these species and the yields of some other fibre-producing animals.

TEXTILE 6.3 Yields of wool, textile and other fibres (normal range)

Animal	Annual yield per head (kg)
Sheep	1–9
Alpaca	1·5–5
Llama	1·5–3·5
Vicuna	0·18–0·32
Huanaco	1–2
Musk oxen	2·3–2·7
Goat (Cashmere)	0·23
Rabbit (Angora)	0·23
Camel	2·3–4·5

The proportion of the wool or hair removed may vary with the method employed and several kinds of wastage can be identified, but the efficiency of harvesting is probably not of great practical importance in the sheep industry. By and large, if rather less wool is shorn and a slightly greater length of fibre left behind, the next wool-clip will be slightly greater.

Another important non-food product is leather. This, of course, involves the slaughter of the animal and the efficiency of harvesting is thus largely a matter of care in removal of the hide and in its processing. The yields obtainable from different animal species are indicated in Table 6.4.

TABLE 6·4 Yields of leather (source: Raistrick, 1973)

Animal species	Yield per head (ft^2)
Buffalo	35–50
Camel	30
Cattle	35–50
Ass	15
Horse	35–50
Goat	4·5
Pig	10–12
Reindeer	20
Sheep	6
Yak	30
Elephant	100
Hippopotamus	60
Eland	40
Wildebeest	24
Kudu	20
Impala	8
Kangaroo	10
Ostrich	12

In general, the more controlled the agricultural system, the more efficient harvesting can be. The extremes can be observed where the product represents progeny that could be used for breeding and thus further production. In a well-controlled dairy cattle enterprise, for example, the number of female calves that must be retained for replacement of the adult population can be precisely calculated and as precisely effected. In a wild population of animals, on the other hand, the optimum numbers to harvest and leave in the population are much more difficult to calculate and to arrange. For example, it has been estimated that the allowable annual harvest from wild African elephants is 4·8% of the biomass (Petrides and Swank, 1965) or only 2·6% of all elephants if harvesting is restricted to adults.

In freshwater fish production, unless ponds are completely emptied periodically, harvesting has to try and separate those fish that it is desired to catch from those it is proposed to leave for breeding and from those that it is not desired to catch yet.

An intermediate situation, allowing more controlled harvesting, depends upon "batching" of some kind, so that all the animals in one population are breeding, or growing, or ready to be harvested together. This procedure often raises problems of space allocation, because a batch of animals consists of individuals all getting bigger at the same time. This difficulty is most acute if the space allotted serves both as living and feeding area and as the area for growing the food (as in grazing situations).

In a wild population, there is a proportion of the population that can be harvested without detriment to the maintenance of population size, but, as noted in the case of African elephants, this proportion varies with the particular fraction removed. Thus, in wild sheep, a quite different proportion of males and females can be removed. Any expression, for example of the percentage of energy or protein that can be removed, that ignores these considerations, must be of doubtful validity. Any single expression of the maximum that can be removed, therefore, would most usefully apply to removal of the least damaging fraction. The latter depends upon what other predation is taking place and many agricultural systems have as a major aim early in their development the reduction of predation from any source other than the agriculturalist.

It would seem as though the choice of animal production systems ought to be based on selecting a species that has evolved to meet very heavy predation and then to substitute agricultural harvesting for predation. This might apply also to the nature and timing of the harvest. Thus, a bird might be domesticated because it had evolved to meet heavy predation at the egg stage. Clearly, however, it could be used either for egg production or by removing the eggs as starting points for another enterprise. One of the main arguments in favour of egg production from marine turtles (see Chapter 5), is that a very high proportion of the eggs laid would normally be lost anyway.

4. Indirect Harvesting of Secondary Production

This is undoubtedly the least important category of agricultural harvesting. The reason for this is that carnivores have not generally been used agriculturally, mainly because they are not efficient users of land for the production of meat, hides, milk or eggs.

Carnivores require large areas of land to support and feed their prey and the populations from which the prey are derived. Only a small proportion of these populations can be removed as prey, without reducing the growth of the populations themselves. It is fortunate, from this point of view, that the old and the diseased are usually easier to catch and also that their removal is least damaging to the population.

Predators may also be fierce and difficult to manage agriculturally, as well as being equally likely to prey on the agricultural manager. If they are large and fierce, they are less likely to be preyed upon themselves and are very unlikely to have evolved a high reproductive rate. If they are relatively small and manageable, they may be used for hunting. Cheetahs have been used for this purpose and so have hawks; the use of cormorants for fishing is very similar.

Agricultural use of carnivores really requires that the animals are employed as foragers, over a large area of land or sea, unaccompanied by man (the hunting is then by the animal but not by man), and that they return to an agricultural base where they can be harvested.

One example of this is the use of sea birds by the people of the St. Kilda group of Scottish Islands (Steel, 1965). The latter are isolated and difficult to fish from because of the dangerous sea and the nature of the cliffs. Some sheep (e.g. Soays) were kept but the main animal enterprise was the harvesting of sea birds. Indeed, fowling for gannet, fulmar and puffin was a major activity of the islanders and these birds were considered to represent the equivalent of the Laplander's reindeer. Gannets were taken off the nests at night, the "sentry bird" having been killed first. Fulmar oil (*ca.* $\frac{1}{2}$ pt per bird) was used in lamps, feathers were used and sold, both meat and eggs were eaten and unused parts of the bird were used as manure. It was estimated that about 20,000 gannets were being killed annually in the eighteenth century: this may have represented some 10% of the estimated 200,000 gannets on Hirta (also estimated to be consuming annually some 306,000 barrels of fish!).

This appears to be an example where a settled community could rely on finding the carnivorous birds at the same place and harvesting them at appropriate times in a controlled fashion. In these particular circumstances, it represented a more efficient total process than catching and eating the fish directly and had the advantage of providing products (such as feathers) that fishing could not provide.

The same kind of process can be visualized on land (based on vultures?) but seems most likely to be associated with harvesting the sea. Turtle farming is essentially similar, where a degree of intervention is employed. The cultivation of eel ponds is a curious blend of these activities and fishing.

References

Aylward, F. and Hudson, B. J. F. (1973). *In* "The Biological Efficiency of Protein Production". (J. G. W. Jones, ed.), Chapter 17. Cambridge University Press.

Blaxter, K. L. (1962). "The Energy Metabolism of Ruminants". Hutchinson, London.

Blaxter, K. L., Wainman, F. W. and Wilson, R. S. (1961). *Anim. Prod.* **3** (1), 51-61.

Butler, C. G. (1954). "The World of the Honeybee". Collins, London.

Free, J. B. and Butler, C. G. (1968). "Bumblebees" (2nd edition). Collins, London.

FAO (1971). Report of an ad hoc consultation on the value of non-protein nitrogen for ruminants consuming poor herbages. Kampala, Uganda, 29 June-3 July, 1971.

Gerrard, F. (1971). "Meat Technology" (4th edition). Leonard Hill, London.

Gibb, M. J. and Treacher, T. T. (1974). Personal communication.

Heinrich, B. (1973). *Scient. Am.* **228** (4), 97-102.

Johnson, W. (ed.) (1931). "Journals of Gilbert White". Routledge, London.

Lampila, M. (1972). *Wld Rev. Anim. Prod.* **8** (3), 28-36.

MAFF (1971). "Commercial Rabbit Production". Bull 50, MAFF London.

McMeekan, C. P. (1940). *J. agric. Sci., Camb.* **30**, 387-436.

Moulton, C. R., Trowbridge, P. F. and Haigh, L. D. (1922). *Res. Bull. Mo. agric. Exp. Stn* **55**.

Pálsson, H. and Vergés, J. B. (1952). *J. agric. Sci., Camb.* **42**, 1-92.

Petrides, G. A. and Swank, W. G. (1965). *Proc. 9th Int. Grassl. Congr.* **1**, 831-841.

Raistrick, A. S. (1973). Personal communication.

Rodriguez, J. M. and Hodgson, J. (1974). *Proc. 12th Int. Grassl. Congr. Moscow* (In press).

Spedding, C. R. W. (1970). "Sheep Production and Grazing Management" (2nd edition). Bailliere, Tindall and Cassell, London.

Spedding, C. R. W. and Diekmahns, E. C. (eds). (1972). "Grasses and Legumes in British Agriculture". Bull. 49. Commonw. Bur. Past. Fld. Crops, Hurley, p. 511.

Steel, T. (1965). "The Life and Death of St. Kilda". The National Trust for Scotland.

Treacher, T. T. and Hodgson, J. (1970). Unpublished data.

U.S.D.I. (1963). "Trout Feeds and Feeding". Circular 159. U.S. Dept. of Int., Fish and Wildlife Service.

7 Losses and Wastage in Production Systems

Losses can occur *from* systems of all kinds, but terms such as "wastage" only make sense in relation to a purposeful system in which there is a desired output. It is not very helpful, however, to regard everything as lost or wasted that does not appear in the final product. The desired output, of meat or milk, for example, cannot conceivably contain all the energy used in its production. There are, in other words, production costs of many kinds, and at least a proportion of those incurred are inevitable and cannot sensibly be described as losses or wastage.

Wastage suggests losses that could have been avoided. These may be of two kinds: first, losses of product or intermediate product material that could have been harvested or converted; and secondly, production costs that could have been avoided. Thus, if potatoes already grown are not harvested, or are eaten by pests or rot during storage, we are clearly dealing with losses that could have been avoided. However, avoiding them might have cost more than they were worth, so we may be obliged to tolerate some losses for economic reasons. Avoidable production costs may have to be tolerated for similar reasons. For instance, very high growth rates in meat-producing animals lead to a higher proportion of the total food being converted into product. An unnecessarily high proportion of food being used for maintenance is thus avoided. But if the higher animal growth rates require more expensive foods, it may prove unprofitable to use them, even though they may be more efficiently used in terms of food conversion.

Loss and wastage, therefore, do not apply only to substances already formed, but include at least some aspects of reduced performance. This is important because, at first sight, it might seem sensible to exclude the latter. If we imagine one tree on an area of land that could support two, it does not seem very useful to describe the missing tree as a loss, unless it had been there once and had been removed. Incident light might nevertheless be considered wasted for the lack of photosynthetic material. Similarly, ungrazed vegetation, due to lack of stock, represents wasted vegetation rather than lost animal production. There are thus straightforward losses of what has already been produced and wasted resources that represent a loss of potential production.

In some circumstances these are described as "direct" and "indirect" losses. When a sheep is infected with internal parasites, for example, there may be small direct losses of substance (gut wall, blood, etc. in the case of stomach and intestinal nematodes) but, in addition, large losses of potential growth, due to reduced appetite and lowered digestive efficiency (see Spedding, 1970). As a consequence of a reduced growth rate, there is then a loss of nutrients because the proportion required for maintenance is increased.

The most satisfactory definition of "losses" then appears to be "diversion of a greater than optimum flow of matter (however described) from the main production process". This means that it is necessary to state the product being considered and what sort of losses (of calcium, energy, dry matter) are being described; it is also essential to attach a meaning to the word "optimum" in this context. For any of the material flows in a production system to have an optimum rate, it is necessary to describe the production required from the system, per unit of some or all of the resources.

None of this would matter greatly, if it were not for the fact that some deceptively obvious losses often turn out to be unavoidable or avoidable only with adverse consequences to the system as a whole. This will be clearer if the main sources of loss in production systems are examined. These may be identified from a diagrammatic representation of any particular kind of system, such as that of goat production from browsing shown in Fig. 7.1. It will be seen that there is an opportunity for loss to occur throughout the production process. This process is continuous, although phases can be detected and points determined at which what has been happening may usefully be summarized. Losses may usefully be summarized at the same points, but this should not obscure the continuous nature of the wastage processes.

In Fig. 7.1 the units are not specified and it is implied that they are the same throughout: this is the simplest situation but frequently one element may be used and lost in the production of another. Furthermore, if the unit chosen had

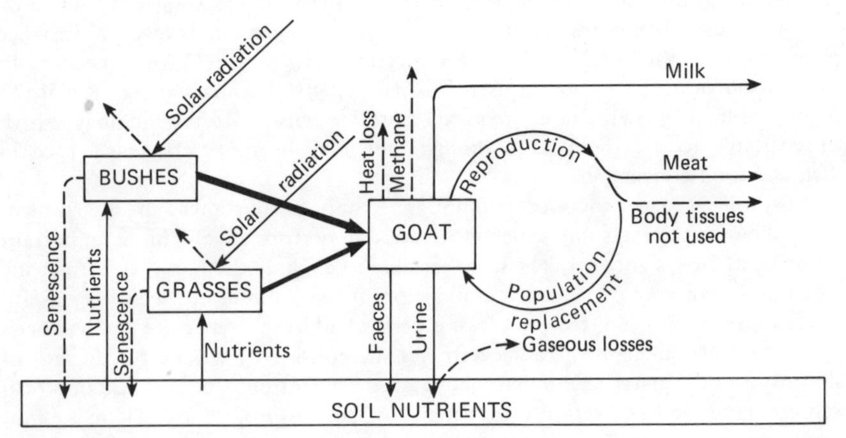

FIGURE 7.1 The main material flows and losses (– – – –) in a goat browsing situation.

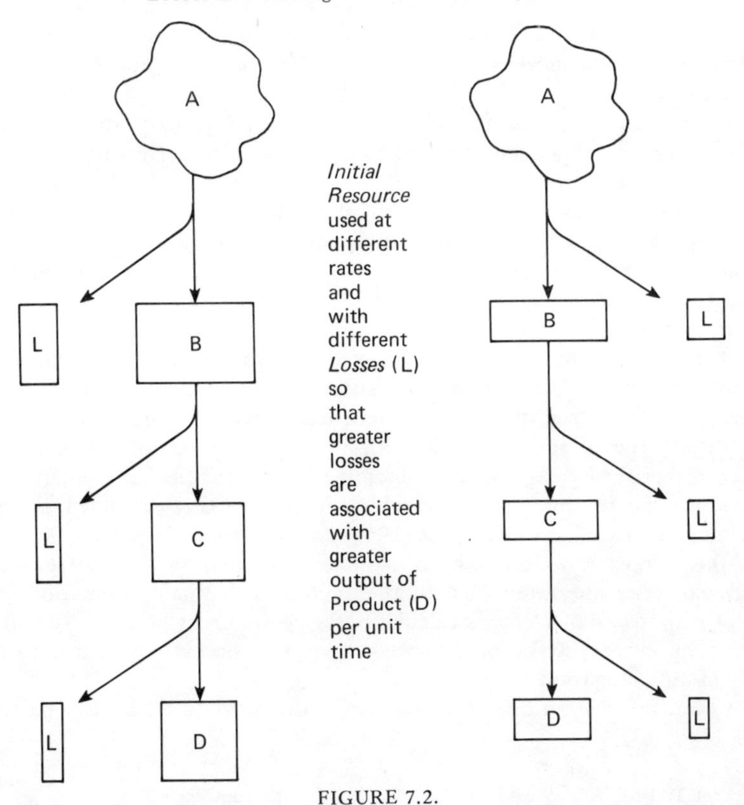

FIGURE 7.2.

been energy, no recycling would have occurred and all "losses" would have left the system. In the case of nitrogen, however, the rate of loss at one point has to be balanced by the amount recycled and not lost to the system as a whole.

Furthermore, it is possible for productivity of the whole system to depend upon high rates of resource use that involve high losses at that part of the system. Consider the following two situations, where simple systems are represented by the transfer of material from $A \rightarrow B \rightarrow C \rightarrow D$ and the size of the boxes is proportional to the amount of material (see Fig. 7.2). In this particular case, the initial resource (A) must be regarded as outside the system and the greater losses at all stages are associated with a greater input of resource as well as greater output of product. Any attempt to reduce one of the losses in such a system may result in a decrease in the rate of the production process.

Thus, just as a "benefit" in one part of a system has to be evaluated in terms of its effect on the whole, so it is with losses; and the main reasons for this are:

(a) that losses at one stage may be recycled within the system, and
(b) that the magnitudes of the losses may be positively related to the rates of production.

The second situation is well illustrated by the use of resources not directly involved in the biological production process and certainly not included in the final product.

Since resources are commonly not available in exactly the proportions required, some are generally limiting and others not. The introduction of additional quantities of a limiting substance can often have a disproportionate effect on productivity. Indeed, the total absence of a vital element results in no production at all and even small supplies of this element may result in substantial production, especially if it is a substance that is recycled within the system.

This applies to all elements involved in the total process, whether they form part of the product or not. But very large quantities of additional inputs may be used in modern agriculture in order to improve performance or to reduce losses. This is particularly true of energy. Labour has always been used in this way: in the earliest forms of agriculture—as with hunting and food-collecting—considerable energy might be expended in obtaining food which had to contain, at the very least, a greater quantity of available energy than that expended, as well as containing other vital materials (Lee, 1969; Lawton, 1973; Spedding, 1974).

In this connection, it has been estimated (Weiner, 1964) that sub-human primate societies use about 70% of the energy they obtain from food in the search for more food. Weiner's calculations suggested that between 30 and 40% of the total energy intake of a family engaged in simple agriculture was used simply in obtaining food.

However, reliance on human labour limits the power available at a particular

TABLE 7.1 Work capacity of animals (Data from Epstein, 1969)

Animal species	Load capacity	
	Draught	*Carrying*
Humpless cattle of China (bull)	600–1000 kg (depending on hilliness of road)	
Yak		120–150 kg for 13–16 km/day
Mongolian pony	800–1000 kg (over 40–60 km/day)	80–100 kg (over 40–60 km/day)
Szechwan ponies	600–800 kg 40–60 km/day	100 kg over 30–40 km/day
Siumi ass		80 kg
Turkestan one-humped camel	500 kg	200–250 kg 35 km/day
Reindeer	sledge 45–135 kg 32–40 km/day	40 kg for 4–5 h

time and particular times are often critical for sowing, cultivating or harvesting, whilst weather conditions are suitable. Some of these difficulties can be met by cooperative efforts and this is still true in both primitive and highly developed forms of agriculture.

The use of other animals for traction and transport was a further development in this direction. A horse or buffalo has to be fed all the year round but is then able to supply a concentration of power, not previously available (see Table 7.1 for examples of the work capacity of animals). The use of such power might increase the efficiency of the original production process, because of the greater area of land that one man can cultivate, for example, or it may make a form of agriculture possible that was not possible for an unaided man.

None of this would be beneficial, of course, if the food required by the horse or buffalo exceeded the additional production due to their use (see Table 7.2 for the energetic efficiency of a working horse). Figure 7.3 illustrates an example of

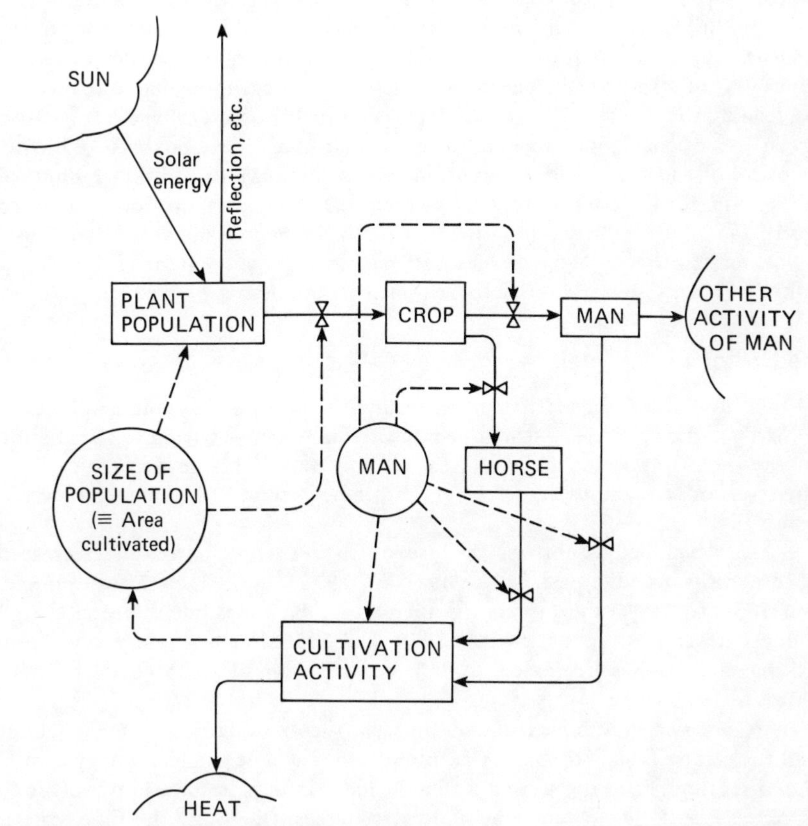

FIGURE 7.3 Diagrammatic representation of the role of a horse in the energy flow of crop production. Boxes represent energy levels and continuous lines represent flows of energy. Circles and broken lines relate to influences.

TABLE 7.2 Energy expenditure of a 1500 lb (680 kg) work horse (Source: Brody, 1945)

Hours work in 24	12	10	8	6	4	2	0
Hours rest in 24	12	14	16	18	20	22	24
Work accomplished (cal)	7700	6420	5140	3850	2570	1280	0·0
Energy expended (cal)	46 600	41 300	35 900	30 600	25 200	19 900	14 600
All day energetic efficiency (%)	16·5	15·5	14·3	12·6	10·2	6·4	0·0

this kind of situation. The point to note is that none of the food fed to the horse, nor the horse itself, actually enters into the main production process. Indeed, it would be possible to regard the food used by the horse as a loss that should be reduced or eliminated. In passing, it may be noted that, unless excessive quantities of food are being fed to the horse, it is better to feed it properly or to dispose of it. The worst outcome results from trying to reduce the loss, leading to a horse which still eats but is incapable of anything else.

The use of machinery to substitute for man- and horse-power is a further extension of the same process and often results in enormous additional inputs of energy. Since the latter can easily exceed the energy in the food harvested (Black, 1971), it is clearly a situation worth examining in some detail.

The main difference in the case of machinery is the extent to which the additional energy source is fossil fuel of one sort or another.

The Use of Fossil Fuels

Fossil fuels are those that have resulted from photosynthetic processes (primarily): they represent stored energy originally derived from solar radiation and the most important are coal, gas, oil and petrol. The magnitude of these stores is not known with certainty and new reserves are being discovered continually.

In the developed countries, the use of such energy sources has increased phenomenally over the past 120 years (Watt, 1973). In the U.S.A., for example, from 1850 to 1970, the population increased 8.82 times but the total energy production was 203 times as great in 1970 as in 1850. Furthermore, energy from working animals has decreased from a contribution of *ca.* 50% in 1850 to 0·008% in 1970.

There are two very powerful and apparently contradictory arguments about fossil fuels. The first is that since eventually they will be exhausted they should be used sparingly, and the second is that we have, in fact, become better off with the passage of time, in terms of known resources. Obviously the time scale is vital to this discussion and the first argument would be held to be a longer-term one, and the second a relatively short-term one.

No-one argues that fossil fuels should not be used at all (although there are different views as to what they should be used for), so opinions differ chiefly on the rate at which it is sensible to use them. The possible reasons why the rate might be important are chiefly three:

(a) the rate of use is a determinant of the rate at which any undesirable consequences occur (pollution, for example);
(b) the faster the rate of use, the more dependent we may be considered to be upon them (the undesirability of this situation rests on the idea that we could suddenly turn out to be wrong in our estimates of the reserves, or that undesirable consequences might reach intolerable levels faster than expected);
(c) the more quickly reserves are used, the sooner they will be exhausted and the less time there will be for substitutes to be found.

Broadly speaking, these arguments apply to all non-renewable resources but it may be argued that substitutes can be found (e.g. plastics) for all except solar energy (current or past). This is not quite true, of course, since atomic energy, that derived from water, and tidal energy are already used to a limited extent. Furthermore, only a very small proportion of the current incident solar energy is used, even for photosynthesis, and geothermal energy has yet to be tapped (via steam, water or hot rocks) on any scale.

The first essential in such matters is to establish the facts, as far as they can be ascertained: but there are bound to be large errors in current estimates of reserves and the arguments are better based on the other issues. After all, if the resources are to be used, they will one day be exhausted, so this prospect has to be faced.

The efficiency with which such resources are used is of great importance, of course, especially in relation to energy. Minerals may be inefficiently used but are entirely recoverable, although at some cost, whether they are used efficiently or not. Energy, on the other hand, is lost when it is used and may be completely lost without being utilized.

Some estimates have been made (Black, 1971; Lawton, 1973; Leach, 1973) of the quantities of fossil fuel energy now used in association with highly mechanized agricultural processes. Table 7.3 illustrates two examples.

It is easy enough to add in to the calculation such items as tractor fuel and electricity used in milking, but it is much more difficult to compute the energy costs of making and servicing the tractor and the milking equipment. In fact, it is difficult to know where to stop, since the machinery used in making the tractor was itself probably made with the aid of other machinery. There is an urgent need for the construction of energy budgets on which sensible planning can be based, of the kind calculated by Leach (1973), including all the energy costs incurred before and after the production process itself (Odum, 1971).

Leach (1973) has calculated energy ratios

$$\left(\frac{\text{energy output}}{\text{additional energy inputs}}\right)$$

TABLE 7·3 Support energy usage in agricultural production (Compiled by Jean M. Walsingham)

		Lamb[c]	Eggs[d]
Support energy inputs	Upstream[a] (Mcal)	71 815	12 695
	On the farm (Mcal)	30 183	115 527
	Downstream[b] (Mcal)	6816	[e]
	Sub-total (Mcal)	108 814	128 222
Labour inputs	Upstream (h)	38	102
	On the farm (h)	1745	1889
	Downstream (h)	530	[e]
	Total	2313	1991
Energy content of the food required to support these hours of work (Mcal)		934	804
Total support energy inputs to the production process (Mcal)		109 748	129 026
(MJ)		458 747	539 329
Energy output	Main product (Mcal)	23 026 (Meat)	16 863 (Eggs)
	By-product (Mcal)	17 040 (Wool)	3480 (Cull carcasses)
	Output total (MJ)	167 476	85 034
Ratio	$\frac{\text{Energy output in products}}{\text{Support energy input}}$	0·37	0·16

[a] used in the manufacture or production of inputs to farm enterprises
[b] used in the utilization of farm products (processing, packaging, delivery)
[c] based on a flock of 300 ewes } on arable farms
[d] based on a 1000 bird battery unit } on arable farms
[e] data needed but not available when table was constructed

for various farming systems in the U.K. The values vary from 1·8 to 2·2 for cereals and just over 1·0 for potatoes (at the farm gate) to 0·16 for battery-produced eggs and 0·11 for broiler hens. Milk (0·33) and sugar beet in the shops (0·49) are intermediate. However, fish from freezer trawlers (based on fuel only and to the dockside) have a value of 0·05.

In spite of these variations, the return in terms of the value of the output per unit of fossil fuel input appears to be similar over a wide range of systems. Furthermore, the energy consumption of agriculture is small relative to that of industry.

Even so, agricultural production often depends upon industrial processes and the extent to which this is so frequently determines the energy subsidy that an agricultural system requires. Slesser (1973) has calculated this subsidy for 131 food-producing systems and found great variation between them. He points out that assessment of the energy requirement of a food production process is one way of assessing the long-term viability of such a process and may be used in food policy planning.

Although it has been argued that energy is in a rather special category, some other resources are similar in some respects.

A great deal of water is used, even in non-irrigated crop production (see Table 7.4) and modern industry, incuding dairying, for example, uses vast quantities: the treatment of this water for re-use may require substantial quantities of energy. Water is limited in total quantity and many countries are now using appreciable proportions of the total available to them. In arid areas, where irrigation has to be used, quite extraordinary quantities of water may have to be applied. For example, it has been calculated that about 1000 t of water (evaporated, transpired or used directly) are required in irrigated lands to produce 1 t of sugar or maize (Dasmann, 1972). In the U.K., a typical household uses some 36 gal a day and factory usage is enormous. It has been estimated that a pint of beer involves 3 gal of water use; a ton of steel, 44 000 gal; a gallon of petrol, 70 gal; a small car 100 000 gal; a daily paper, 42 gal per copy; and a tyre, 42 000 gal (Long, 1973).

TABLE 7.4 Water requirement of plants
(after Kipps, 1970, quoted from Hildreth *et al.*, 1941)

g of water used per g of D.M. produced

Sorghum	322	Soybeans	744
Maize	368	Sweet clover	770
Wheat	513	Vetch	794
Barley	534	Alfalfa	831
Oats	597	Potatoes	636
Rice	710	Cotton	646
Flax	905	Sugar beet	397

There are many reasons for trying to avoid inefficiency in the use of resources, especially if these are scarce, and it must be remembered that resources are not just something that production processes start with, they are involved at all stages. Furthermore, some of them cannot be turned on and off as required: this is particularly true of labour, which may have to be represented on a farm by one or more full-time men and cannot vary daily or by fractions.

There are also losses of product, both final and intermediate.

Production and Harvesting Losses

In general, these are either of plant or animal material already formed but not used in the main process.

Plant losses may occur during growth or as a result of failure to harvest. As mentioned before, it cannot be assumed, in perennial crops, that all that is usable may be harvested at one time without any effect on subsequent production. Table 7.5 shows what proportion is regarded as harvestable for a selection of major crops.

TABLE 7.5 Approximate proportion of crop that is represented in harvestable yield (Yield D.M. as % of above-ground plant D.M.)

		References
Wheat	54–60	Bland, 1971
Perennial ryegrass (at one harvest)	63	Smith *et al.*, 1971
Maize	42	Bland, 1971
Field peas	50	PGRO Ann. Rep. 1972
Lettuce	50–90	Holmes, 1974
Brussels sprouts	30	Holmes, 1974

Where plant material comes from annual crops, the only limits to what is harvested are practical and economic: the combined effect of costs and market preferences, however, may result in a very small proportion actually harvested of the total plant available at harvest time. In the case of several crops (e.g. barley and peas) a large proportion of the plant is harvested initially but the same machine discards the unwanted straw and haulm.

In root crops, the desire to avoid harvesting lumps of soil may lead to failure to harvest all the roots. Studies of such losses in harvesting sugar beet (Maughan, 1972) showed that they varied with weather conditions, power take-off and forward speeds, but the seasonal average was 1·17 tons/acre or nearly 7%. Comparisons of potato harvesting machinery (MAFF/ADAS/PMB, 1972) showed that most machines left 0·5 ton/acre in the soil and damaged up to 5% of the tubers collected.

Losses in the U.K. in harvesting grain have been found to be in the range 80-93 lb/acre for wheat and 113-184 lb/acre for barley (MAFF, 1971).

Of course, losses do not end with the harvested grain. Before wheat grain appears in foodstuffs for man, for example, it is estimated that the milling process removes about 50% of the protein and 25% of the digestible energy (Hodgson, 1971). Since only part of the wheat plant is in the grain, only about 35% of the protein and about 40% of the digestible energy in the wheat crop is available for human consumption.

It is worth emphasizing at this point the conclusion of Leach (1973) that cereal straw, regarded as a disposal problem at present on U.K. farms, could be harnessed to provide the additional energy requirements of the farms that produce it, at relatively little energy cost, to the point that they could become almost self-sufficient in this respect.

Similar considerations apply to animal production. Harvesting in this context also includes both:

(a) the proportion of the population that can sensibly be removed, at whatever intervals are appropriate, without decreasing the rate of production and
(b) the proportion of what is harvested that is actually used.

The latter is difficult to express, because of the use of by-products, but it can be a very important matter.

Table 7.6 shows how this proportion is commonly expressed for meat-producing animals. But the carcass of all animals contains inedible fractions

TABLE 7.6 Dressing-out or killing-out percentage (K.O.%) in animals: i.e. the proportion of carcass in the whole body

	Approx. K.O.%
Beef cattle	55–60
Lambs	45–50
Pigs	70–75
Rabbits	45–52
Poultry	68–72

(such as bone and gristle) and the non-carcass parts of the animal are sometimes edible. On the other hand, it would be quite possible to select an animal species that was biologically efficient, in the sense that it incorporated into its body a high proportion of what it consumed, and yet resulted in a small edible yield simply because we, by choice or custom, only consumed a special fraction of the body tissue produced.

Losses during Growth

This is a large and important category, both in crop and animal production (see Table 7.7). It includes the main effects of pests, parasites and diseases, in so far as they divert to their own ends what has already been produced. As argued earlier, it is more difficult to regard as a "loss" the prevention of something being produced. In practice, of course, these two processes may be difficult to separate.

TABLE 7.7 % of global losses preharvest due to weeds, insects or diseases (after Cramer, 1967)

	Crop		
Cause of loss	Maize	Wheat	Rice
Weeds	37	40	23
Insects	36	21	58
Diseases	27	39	19
Total loss (L) preharvest (million tonnes)	121	86	207
Harvested crop (H) (million tonnes)	218	266	232
$\frac{L}{L + H} \times 100$	35·7	24·4	47·1

For example, the larvae of the crane fly (family: Tipulidae) feed on the roots of pasture grasses. There is thus a loss of plant tissue but not, as it happens, tissue that would ever have been harvested. Nevertheless, if the larvae are sufficiently numerous, they may consume so much root material that the growth of the harvestable upper parts is substantially reduced. Just how much root consumption has to occur before this happens will vary with such factors as the nutrient and water supplies: in short, with how much root the plant needs.

In animals, the tapeworm (*Moniezia expansa*) of sheep lies in the alimentary tract removing relatively small amounts of nutrient that are then "lost" to the sheep. If very numerous, such tapeworms may cause interference with the digestive functions and affect animal growth but, in a sense, they are more similar to competition for food by other sheep than they are to the roundworms (such as *Haemonchus contortus*) that suck blood from the wall of the abomasum and thus remove tissues already formed (see Lapage, 1956, and Soulsby, 1965, for further details).

It may be that in animals the most serious effects are those on function, perhaps because it is not possible to remove substantial amounts of tissue from living animals without causing death. Losses of this kind, by predation, vary greatly with the degree of control of the agricultural enterprise and may be large in some parts of the world.

In plants, on the other hand, life can continue in spite of heavy depredations and the distinction between pests and parasites is less obvious. An aphid is a sap-sucking pest of plants, for example, but a blood-sucking louse (*Linognathus ovillus*) is an external parasite of sheep. Of course, it is also true that the blood-sucking mosquito is not regarded as a parasite, because it lives part of its life away from its host animal (as does the parasitic liver fluke, *Fasciola hepatica*, however).

The numbers and variety of pests and parasites of crops and animals are very large indeed; no part of an organism is entirely free from the risk of such attack but, for obvious reasons, the less vital tissues tend to suffer most. Indeed, it is often suggested that there is generally a balance between plants and animals and their pests and parasites and that problems only arise when this balance is disturbed, as, for example, by over-intensive agriculture.

Now, of course, the same argument could have been applied in the past (and probably was) to man and his parasites, but this does not make such a balance necessarily desirable, even if it can be achieved or maintained. Evolutionary arguments really concern the balance of species and numbers may fluctuate widely from time to time. It is rare for a constant state to be achieved and much more likely that "balance" will be represented by a tendency for an equilibrium situation to be restored after each disturbance. In any event, there may be no great merit in a particular "balanced" state and it may be desirable to shift to a new state.

There is, nevertheless, plenty of room for discussion as to *what* state and how it should be achieved and maintained.

The use of pesticides, for example, may prove a bad way to achieve a given state (see Mellanby, 1967):

(a) because of pollution problems, and
(b) because frequent changes of pesticide may be needed to maintain it, in the face of the development (or survival) of resistant pest populations.

Estimates of the crop losses that would be sustained if no pesticides were used have to be treated with great caution. There are two main dangers of statistical exaggeration. First, the yield of an infested crop may be compared with a good, rather than average, yield; secondly, losses from different sources may be added. In apples, for example, it has been pointed out that 60% may be damaged by insects, 80% by disease and 10% lost due to weeds, a total of 150%; nevertheless, the loss without pesticides has been genuinely assessed at a figure as high as 90-95% (Pimental, 1973).

The area treated with pesticides is now substantial. In the U.S.A., for example, it is reckoned that 81% of the area growing tobacco and 97% growing citrus is treated with insecticide (but only 0·5% of the pasture), 59% of the area growing potatoes is treated with herbicides (but only 2% of the tobacco area), and 35% of the area growing peanuts is treated with fungicide (but only 2% of the cotton-growing area).

However, there are substantial worries both on the grounds of pollution and the possibility that pesticides represent too crude a weapon to be effective. This leads to the advocacy of "biological control".

Biological Control

This simply implies the use of one biological agent to control another and does not exclude the combined use of such agents with, for example, chemical methods; this is sometimes called integrated control.

There have been several more or less spectacular successes and failures in biological control (van den Bosch, 1971) but it seems likely that there are relatively few situations where the use of one simple method will prove completely successful on its own.

It has been calculated by van den Bosch that some 81 fully or substantially successful biological control programmes have been carried out along classical lines since the first success nearly 100 years ago in California (*Rodolia cardinalis* vs. *Icerya purchasi*). The most striking successes are listed in Table 7.8.

Some undesirable pests and diseases can be eliminated but many are likely to remain widespread. It is difficult to imagine, for example, the complete elimination of aphids. In such cases, the aim is to control numbers of pests or the level and extent of disease.

Biological agents that prey on pests, or parasitize them, have to have a supply of food and will themselves die out before their prey, especially if they live on nothing else.

However, as with chemicals, so with predators and parasites, the effect of using them is rarely simple. Chemicals may kill organisms other than the one aimed at: some of those killed may be predators or prey of something else and

TABLE 7.8 Examples of successful biological control against insects (after van den Bosch, 1971)

Insect controlled		Country
Winter moth	*Opheropotera brumata*	Canada
Southern green stink bug	*Nezara viridula*	Hawaii
Olive scale	*Parlatoria oleae*	California
Rhodesgrass scale	*Antonina graminis*	Texas
Walnut aphid	*Chromaphis juglandicola*	California
Alfalfa weevil	*Hypera postica*	Eastern U.S.A.
Carrot aphid	*Cavariella aegopodii*	Australia and Tasmania
San Jose scale	*Quadraspidiotus perniciosi*	Switzerland

the balance of species then changes. Insects introduced to deal with one pest may also eat others or destroy the food of others. There are, of course, some very sophisticated approaches within biological control. The sterile-male technique has been used successfully for the eradication of the screw-worm (*Cochlionyia hominivorax*), a serious parasite of livestock. Males are sterilized by ionizing radiation (in the range 5-10 krad) and then released in overwhelming numbers to compete with normal males, all on a vast scale (Bushland, 1971).

The relevance of this topic to the present discussion lies in the implication that very often a balance of species and numbers will be the desirable outcome and not simply the removal of an undesirable organism. Some losses may thus have to be tolerated, even where they can be avoided, simply because the loss can be controlled, whereas elimination of this loss entirely may result in an increase in other sources of loss that may be less controllable.

Losses in Reproduction

As mentioned in Chapters 4 and 5, rates of reproduction may be very important in the agricultural use of plants and animals.

This is clear for crops in which seeds or fruit are the products and for animals kept for meat. It is also important wherever the productive life of the species is short, whatever the nature of the product, since new individuals are then required at frequent intervals.

If annual plants produce very little seed and new crops depend upon being sown, many plants have to be devoted to the production of the necessary seed. Similarly with animal species, where both reproductive rate and longevity are low, substantial adult populations have to be maintained for breeding purposes. This may also be the case where elaborate cross-breeding programmes are required in order to produce the final product-forming individuals. In some of these situations, however, a spatial distribution of the different breeding components is involved and may be combined with the exploitation of several different environments. This is so, for example, in the stratification of sheep breeds in Britain (see Fig. 7.4).

In general, however, the less energy (or substance) that has to be "wasted" on reproduction, the better. The question of what is "waste" in this context has to be carefully considered. In meat production, for example, it is usually productive to divert as much energy as possible to reproductive rather than maintenance purposes.

In natural ecosystems, on the other hand, it must be wasteful to devote more energy to reproduction than is strictly necessary to maintain the species at a constant level (using the latter as an example of natural purpose: after all, it must be so at some level, otherwise uncontrolled increase would outgrow the food supply). Of course, the reproductive rate needed to maintain numbers varies with degree of predation and losses due to disease and other causes. Some species minimize such losses in other ways, but some depend upon a high

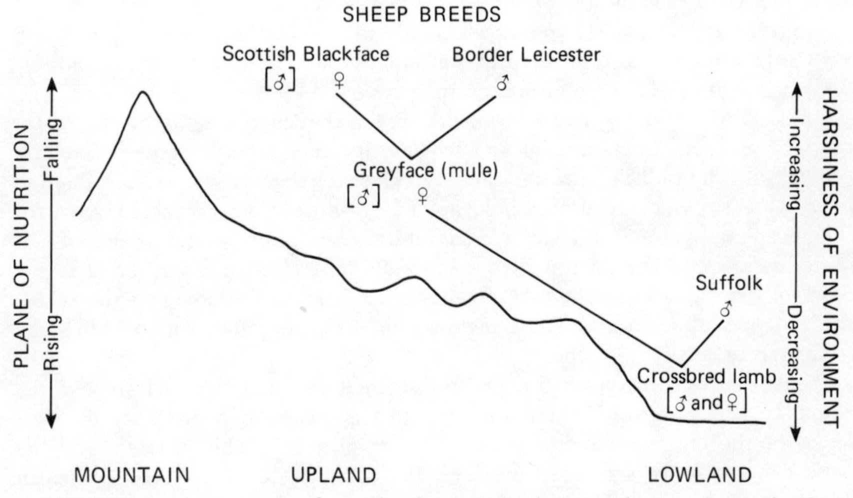

FIGURE 7.4 An example of stratification in U.K. sheep production. Square brackets indicate lambs normally slaughtered for meat, as distinct from being kept for breeding.

TABLE 7.9 The proportion (%) of the annual food intake that is deployed in reproduction by agricultural animals and some plants

Animal	No. of progeny	% of intake
Ewe	1	42
Ewe	2	51
Ewe	3	56
Cow	1	37
Sow	10	34
Sow	20	44
Rabbit	40	59
Hen	160 eggs	26
Hen	240 eggs	35
Hen	350 eggs	45
Plant		**"Reproductive effort"[a] x 100**
Solidago nemoralis		28–52
Solidago rugosa		<1–18
Hardwood species		*ca.* 10
Senecio vulgaris		*ca.* 20[b]
Most grain crops		25–35[c]

[a] This is the ratio of dry weight of reproductive tissue to the total above-ground dry weight, used as an index of resources devoted to reproduction (Gadgil and Solbrig, 1972)
[b] calculated from data given by Harper (1967)
[c] from Ogden (1968)

reproductive rate. In many ways, the latter are the desirable species from an agricultural point of view, provided that the losses can be avoided and the high reproductive rates exploited.

Animals and plants (Harper *et al.*, 1970) differ greatly in the proportion of their total energy that is devoted to reproductive activity (see Table 7.9 and Fig. 4.3), although this is by no means a simple evaluation to make.

In a herd of dairy cows producing milk, the energy used primarily for reproduction can be estimated and will be different from that used for milk production; the bull may be considered to exist for the purpose of reproduction only. The fact that lactation would not occur without parturition may be, for this purpose, ignored. But in a wild rabbit population, lactation is part of reproduction and the energy used to produce milk has to be included in the costs of reproduction. There is no more justification for charging the whole energy costs of maintaining the male rabbit, however, than for including the maintenance costs of the doe.

Thus the view taken of energy transformations in reproduction and the extent to which any of them may be regarded as wasted, depends upon the view taken of the entire biological system and its purposes. This is essentially the same conclusion as is reached whatever the components of a system on which one focuses initially. The value of considering "losses" as separate items is limited and it is usually better to look at the system as a whole, in order to assess the total effect of changes in any part of it.

References

Black, J. N. (1971). *Ann. appl. Biol.* **67**, 272-278.

Bland, B. F. (1971). "Crop Production: Cereals and Legumes". Academic Press, London and New York.

Bosch, R. van den (1971). *A. Rev. Ecol. Syst.* **2**, 45-66.

Brody, S. (1945). "Bioenergetics and Growth". Reinhold, New York.

Bushland, R. C. (1971). *In* "Sterility Principle for Insect Control and Eradication". Proc. Symp., Athens, 1970, 3 IAEA, Vienna.

Cramer, H. H. (1967). "Plant Protection and World Crop Production". Farbenfabriken Bayer AG—Leverkusen.

Dasmann, R. F. (1972). "Planet in Peril?" Penguin Books—UNESCO.

Epstein, H. (1969). "Domestic Animals of China". Commonwealth Agricultural Bureau.

Gadgil, M. and Solbrig, O. T. (1972). *Am. Nat.* **106** (947), 14-31).

Harper, J. L. (1967). *J. Ecol.* **55**, 247-270.

Harper, J. L., Lovell, P. H. and Moore, K. G. (1970). *A. Rev. Ecol. Syst.* **1**, 327-356.

Hildreth, A. C., Magness, J. R. and Mitchell, J. W. (1941). *U.S.D.A. Yearbk.*, 292-307.

Hodgson, R. E. (1971). *J. Dairy Sci.* **54** (3), 442-447.

Holmes, W. (1974). Proc. 21st Easter School of Agric. Sci., Nottingham. (In press)

Kipps, M. S. (1970). "Production of Field Crops" (6th edition). McGraw-Hill, New York.

Lapage, G. (1956). "Monnig's Veterinary Helminthology and Entomology" (4th edition). Williams and Wilkins, Baltimore.

Lawton, J. H. (1973). *In* "Resources and Population" (B. Benjamin, P. R. Cox and J. Peel, eds), pp. 59-76. Academic Press, London and New York.

Leach, G. (1973). *In* "The Man-Food Equation" (A. Burne, ed.). Academic Press, New York and London.

Lee, R. B. (1969). King Bushman subsistence: an input-output analysis. *Bull. natn. Mus. Can.* **230**, 73-94.

Long, E. (1973). *Farmers Weekly*, Jan. 12, p. VIII.

MAFF (1971). "The utilisation and performance of combine grain loss monitors 1971". Ministry of Agriculture, Fisheries and Food, London.

MAFF/ADAS/PMB (1972). Farm mechanisation studies No. 24. Joint investigation by MAFF, ADAS and the Potato Marketing Board.

Maughan, G. L. (1972). *Br. Sugar Beet Rev.* **40** (5), 215-217.

Mellanby, K. (1967). "Pesticides and Pollution". Collins, London.

Odum, H. T. (1971). "Environment, Power and Society". Wiley-Interscience, New York.

Ogden, J. (1968). Ph.D. Thesis, University of Wales.

PGRO (1973). *Pea Growers Res. Org. A. Rep.* 1972.

Pimental, D. (1973). *Environment* **15** (2), 18-31.

Slesser, M. (1973). *J. Sci. Fd Agric.* **24**, 1193-1207.

Smith, A., Arnott, R. A. and Peacock, J. M. (1971). *J. Br. Grassl. Soc.* **26**,, 157.

Soulsby, E. J. L. (1965). "Textbook of Veterinary Clinical Parasitology. Vol. 1, Helminths". Blackwells, Oxford.

Spedding, C. R. W. (1970). "Sheep Production and Grazing Management" (2nd edition). Bailliere, Tindall and Cassell, London.

Spedding, C. R. W. (1974). *Biologist* **21** (2), 62-67.

Watt, K. E. F. (1973). "Principles of Environmental Science". McGraw-Hill, New York.

Weiner, J. S. (1964). *In* "Human Biology". (Harrison, G. A., Weiner, J. S., Tanner, J. M. and Barnicoat, N. A., eds), pp. 401-507. Oxford University Press.

8 Agricultural Ecosystems

In the preceding chapters (Chapters 4-7), component processes have been considered, in sequence and somewhat in isolation. This was accepted as necessary because it is not possible to look at whole systems all the time, but a resynthesis is now appropriate, in order to see how these processes fit together in agricultural ecosystems. The latter are really the same thing as agricultural systems, since these are generally based on living organisms; the exceptions are component agricultural systems concerned with non-biological processing of products or preparation of foodstuffs.

However, the term ecosystems is being deliberately employed in this chapter to emphasize the ecological approach to whole agricultural systems and not merely to their internal structure and functions. The intention here, then, is to discuss the interactions between agricultural ecosystems and the environments within which they operate. These environments include man and the effect of agriculture on people is therefore bound to be included. There are some special relationships between man and the applied biology on which agriculture is based, however, that will be treated separately (in Chapter 9).

The main elements in this ecological view of agricultural systems are:

(1) the provision of greater internal detail;
(2) a longer-term consideration; and
(3) a more complete account of external relations.

1. Internal Detail

There is always a greater content in real agricultural ecosystems than can ever be considered in our pictures of them: this is naturally so, since our models are, by definition, simplified abstractions. The missing detail, if our models are all adequate, cannot be essential to our view(s) of the system, or we would have had to include it.

Perhaps other views are also desirable and *would* require some of this detail, but it is difficult to make any predictions about this. More commonly, perhaps, our existing views are adequate in the short term, but may be less satisfactory over longer periods. After all, some biological processes, such as the growth of trees and elephants, take a long time: over a short period, the increase in size of a sapling tree or a baby elephant may make no noticeable difference to the systems containing them, but this might not remain true over longer periods.

The most general case, where missing detail may become important, is probably where changes in inputs and outputs do not reflect important changes of system content. Clearly, outputs do not necessarily balance inputs, because these may be expressed in different terms and because system content may change.

Particularly for large systems, the output may be greater or less than the input (of any element, including energy), simply because the system itself either contributes to the output or retains some of the input. A tank of water represents a simple example, in which the rates at which water leaves and enters may not be related because the content of the tank may increase or decrease. Theoretically, if the tank were a large enough reservoir, discrepancies of this kind could go on for a very long time. This can generally only happen continuously in one direction over relatively short periods, however, or to a minor extent, or in an alternating fashion, but it may be of considerable biological or economic significance.

Thus many grazing animals lose weight during the winter, when their output of energy exceeds their low inputs of food energy, and gain weight during the spring and summer, when the food supply exceeds their current needs. The same applies to such animals under conditions of severe drought. Survival in many harsh environments depends upon these processes.

Economically, exactly the same kind of processes of weight change may allow a high-yielding dairy cow to produce more milk at the peak of lactation (and thus higher profit) than her capacity for food intake can sustain. There are also changes represented by gradual loss or accumulation of a variety of substances that in time lead to deficiencies or excesses that eventually show up in the outputs of the system. Unfortunately, these changes sometimes reach almost irreversible levels before they are detected and things that are slow to accumulate may also be slow to dissipate or disperse.

Deficiencies sound easier to put right and in the case of minerals this is probably so; but a deficiency of soil fauna or of a beneficial parasite or of insectivorous birds may be much more difficult to correct.

There are also internal changes that make no difference to the functioning of a system, even over a long period of time, provided that the environmental context remains the same. Such systems may therefore tend to become unstable.

A hypothetical example will illustrate the point very simply but there is no shortage of actual examples. Imagine a grazing system in which the pasture consists of two plant species, one of which is a deep-rooted grass and the other a shallow-rooted legume. In combination, we will suppose, they produce the same total amount of nutrients for grazing animals as each plant species does on its own. Thus, if the system changed in favour of either one of these species or another, no change in output would result, provided that the environmental context contained adequate supplies of both water and nitrogen. But if either of these supplies were reduced, one of the single species swards would produce very little, whereas the original mixed pasture would yield at the full level for one of the constituents. The mixed species sward would perform poorly, of course, if both nitrogen and water were in short supply, but would still represent a more stable situation than either of the single species swards.

Changes in plant or animal species are often obvious but may not be apparent to casual observation if they concern very small species, especially if they are in the soil. Soil organisms are very numerous, both as individuals and as species, and may form a complex network of interactions with a marked capacity to adjust to changed circumstances and thus to "buffer" an agricultural ecosystem against environmental change.

It is quite astonishing, for example, what a variety of substances can be dropped on to a soil surface and be quite rapidly decomposed and incorporated within the soil. Senescent plant material is often dealt with at such a rate that no accumulation may be observed at all, dead animals are buried and decomposed by a wide variety of small animals, and faeces and urine are absorbed in remarkable quantities. All such systems can be overloaded but their capacity to operate in a wide range of climatic conditions is impressive. Rates of application of feedlot waste that have been achieved vary from 5 to 300 tons per acre per year (Albin, 1971) but a limit of 72 kg N from feeder cattle has been suggested for 1 ha of land (see also Murphy *et al.*, 1973).

The capacity of even one species of earthworm is notable. So much so that they have been used, not only to convert rabbit excreta to products of value (in this case, potting soil and fishing bait), but also as a method of disposing of cattle dung and dead animals* (Fosgate and Babb, 1972). In the first case (Gaisford, 1972), worms of several species are bred in a 100-doe unit and are estimated to double the returns from such an enterprise. Some strains of worm will apparently produce 2000 offspring a year but they will not breed at temperatures below 40°F (4·4°C). They can be kept beneath the rabbit cages, with benefit to the rabbitry in terms of control of ammonia. In the latter case, *Lumbricus terrestris* was fed entirely on fresh cattle faeces, with added water and lime, and converted faecal dry matter to live earthworm tissue in the ratio of 2 : 1 (or 10 : 1 for raw faeces to live earthworm). The faeces from 1000 dairy cows could thus produce 3174 kg of live earthworm daily, equivalent to 423 kg of animal protein as dried earthworm daily. Dried earthworm meal was found to have 58% protein and 2·8% fat and proved very palatable to cats. The earthworm excreta was found to be equal to greenhouse potting soil for the production of flowering plants but weighed only half as much as normal potting soil.

The activities of earthworms are not normally very noticeable, however. In two woodland populations of *L. terrestris*, the annual means of monthly biomass estimates were 136 and 153 g m^{-2} and production was estimated at 0·33-0·48 g g^{-1} $year^{-1}$ and 0·42 to 0·56 g g^{-1} $year^{-1}$. Tissue production was estimated at 50 to 73 g m^{-2} $year^{-1}$ and 57 to 76 g m^{-2} $year^{-1}$ (Lakhani and Satchell, 1970).

* The cost of disposing of dead cattle presents an increasing problem, estimated at 25 000 tons annually in California with a collection cost of up to $50 per head (Clawson, 1971).

So important are earthworms agriculturally that it has been argued that it would be worthwhile to introduce them as part of every land development programme on soils not already populated (Stockdill and Cossens, 1967). Their beneficial effects include improvement of soil structure, the incorporation of dung, plant residues, fertilizers (and insecticides) and effects on the availability of soil nitrogen (Heath, 1962; Stockdill, 1966; Stockdill and Cossens, 1967). Numbers of earthworms such as *Allolobophora* spp. are greatly influenced by soil temperature and moisture (Gerard, 1967) and numbers have been found to vary with the type of agricultural usage.

In most soils the number of animal and plant species is large and very adaptable indeed. In most cases, it would be possible to remove any one of these species without noticeably affecting the system content or function, because others would expand to fill the gap created. Superficially, it would appear that such a change did not matter and that the species removed was not important. The fact that this may be done with all the species in turn illustrates the limitations of such an argument. The removal of all, or even several, of the species at once might have considerable effects, even if the environment remained constant.

The most important effect, however, might be on the stability of the system, on its capacity to adjust rapidly to environmental change. It is probable, therefore, that a variety of species with overlapping functions will increase stability and that stability will be required to a greater extent in relatively uncontrolled environments. Thus, stable systems may very well be complex, but complexity does not necessarily increase stability. A large number of species, each with its own distinctive role and its own distinctive needs may represent a vulnerable system, simply because "there is more to go wrong". Species with different needs, but overlapping functions, on the other hand, should lead to greater stability.

2. Long-Term Considerations

Stability is one of the most important long-term considerations, simply because longer periods of time allow a greater opportunity for the exposure of systems to extremes of the probable range of environmental change. Time may also be required for some of the consequences of system activity to work through to the outputs or to build up to a significant level.

Thus, agricultural practices that increase vulnerability of soil to wind or water erosion may only actually result in such erosion very gradually. Many undesirable changes flow from long chains of events that take time to develop, especially in the early stages.

Uncontrolled grazing (of goats, for example) may result in very slight overgrazing, the effect of which is to increase the grazing pressure (i.e. the ratio of grazing demand to herbage supply) and thus, in turn, the degree of overgrazing. Deserts are considered to have been caused in this fashion. The need for controlled grazing has been recognized in many parts of the world. Whyte (1964) described the position in India and pointed out that even nomadic and migratory graziers, although clearly operating a rotational system, may severely damage both forest and non-forest types of grazing range. He also drew attention to the resulting, undesirable separation of crop and livestock husbandry, leading ultimately to the impoverishment of both.

In more developed parts of the world, there are greater inputs to agricultural systems and perhaps the more important long-term considerations concern the consequences to the environment and the community external to the agricultural system themselves.

3. External Relations

Every agricultural ecosystem operates within a context that includes all those parts of the world outside that affect or are affected by the system and its activities. Recognition of the extent and nature of this context is one of the most important characteristics of an ecosystem approach.

The extent of the context is not often obvious and the message of the ecologists in recent times might be read as saying that the entire world is a system and an alteration of one component *may* have reverberations throughout it.

There is a kind of parallel to be seen in human relations. It is true that any word said by one person to another (or action taken) may have unforseen consequences, is bound to have some effect and should therefore be pondered most carefully. It also has to be recognized that if this thought renders one speechless or paralysed for fear of doing the wrong thing, one is just as likely to be doing exactly that by inaction. In short, since decisions *have* to be taken on a basis of inadequate information, the message must be accepted seriously, but in spirit.

It would clearly be tragic if we failed to take a good or helpful action because we could not be sure that it would not also have an unhelpful consequence. Similarly, it would be tragic to be left without disease-curing drugs because of the exhaustive nature of the tests imposed to ensure that they could have no conceivable harmful effects in any circumstances. Another very difficult current problem is to balance the need for the heavy (and increased) use of insecticides to increase food production in the tropics, for example, against the possibilities of widespread pollution (Perfect, 1972).

In general, then, reliance has increasingly to be placed on a knowledge of how ecosystems function, so that the results of change can be substantially predicted. This simply involves the development of a systematic way of thinking about the probable consequences of our actions. Being aware of the possible extent of the effects of a system on its context is thus conducive to an open mind but not much of a contribution to reducing the problem to handleable size.

The nature of the context has to be examined in this light. Just as a system has to be represented in the mind by a simplified model, so does the context. Decisions have to be made about the context, in exactly the same way as for the model, leaving out what is irrelevant and keeping the components and interactions that matter sufficiently. This requires an examination of *all* the inputs and outputs of the modelled system and not just of those considered important when looking at the model itself. In fact, it is necessary to consider the original model as a component of a larger system and to model the latter, with rather wider purposes in mind. Since the original model may have been expressed in special terms (e.g. energy or money), it may be necessary to start again and consider the inputs and outputs of other substances in relation to the context.

However, interactions between system and context occur in both directions and it is more logical to consider first the influence of the environment on the system, best illustrated agriculturally by the effect on the choice of system.

The Influence of Environment on the Choice of System

No single factor determines the choice of farming system and it is usually unrealistic to suppose that there is even a single dominant reason for the final choice. Duckham and Masefield (1970) pointed out that the farming systems actually found in any locality are the results of past as well as present decisions by individuals, communities, or government and their agencies. They summarized the sequence of considerations (see Table 8.1) that determine the feasibility, profitability, practicability and preferences that are involved in choice of systems.

Ultimately, it should be possible to consider systematically the resources and constraints of particular situations and to synthesize systems to match. Clearly, environments differ greatly in both available resources and constraints and in the extent to which these can be modified. Systems requirements for resources can also be modified somewhat but all such modifications involve some cost. Some of the possible modifications to resource supply and demand are illustrated in Figs 8.1 and 8.2 for the grazing sheep.

Too little is yet known, however, to make possible the deliberate synthesis of farming systems for specified environments and, as Duckham and Masefield (1970) put it, "the best that we can do today is to accept systematized experience as valid and enhance and evaluate it where possible by quantitative,

TABLE 8.1 Factors affecting the choice of systems (After Duckham and Masefield, 1970)

Ecological factors	Infra-structural features	External economic constraints	Internal operational factors	Personal acceptance
Climatic	Land tenure	Markets	Farm size	Personal preferences
Soil	Water supply	Communications	Labour availability	
Biological	Power supply	Credit availability		

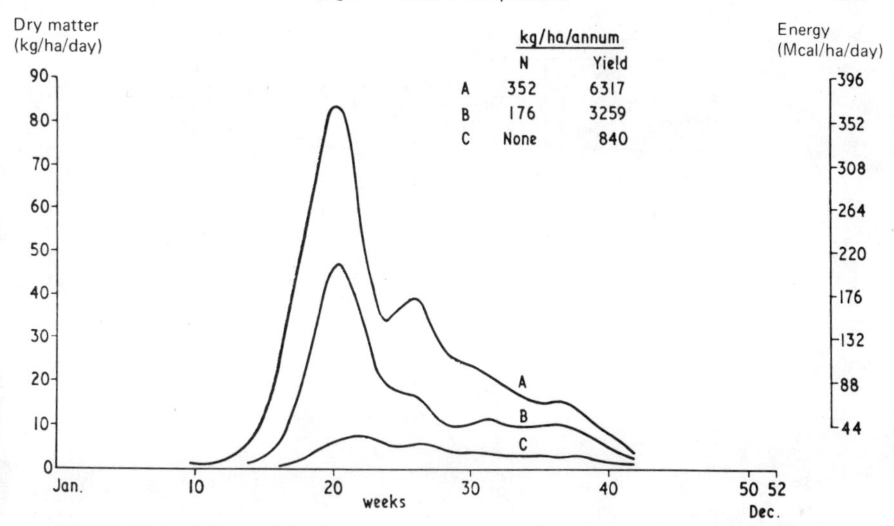

FIGURE 8.1 Effect of fertilizer nitrogen on seasonal pattern of grass production. S23 Perennial Ryegrass. G.R.I. T.311, 1964. Data from A. J. Corrall (from Spedding, 1972).

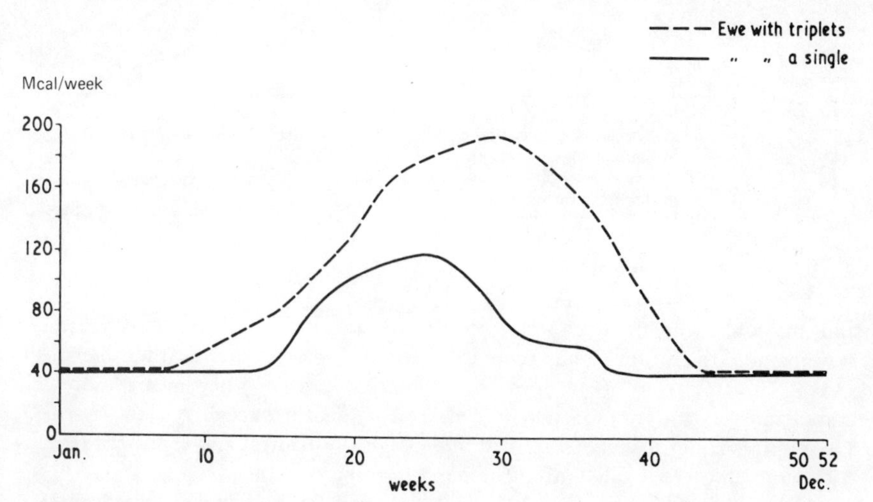

FIGURE 8.2 Mean food requirement curves for a ewe with a single lamb and a ewe with triplets. Data from R. V. Large, G.R.I. H.683 (from Spedding, 1972).

scientific and economic data". These authors found that, in advanced *temperate* regions with a Beckerman Index* greater than about 40 and ample resources per head of population, a well-marked spectrum of farming systems could be identified. Thus there are very extensive grazing systems both in warm, dry areas

* An Index of economic development.

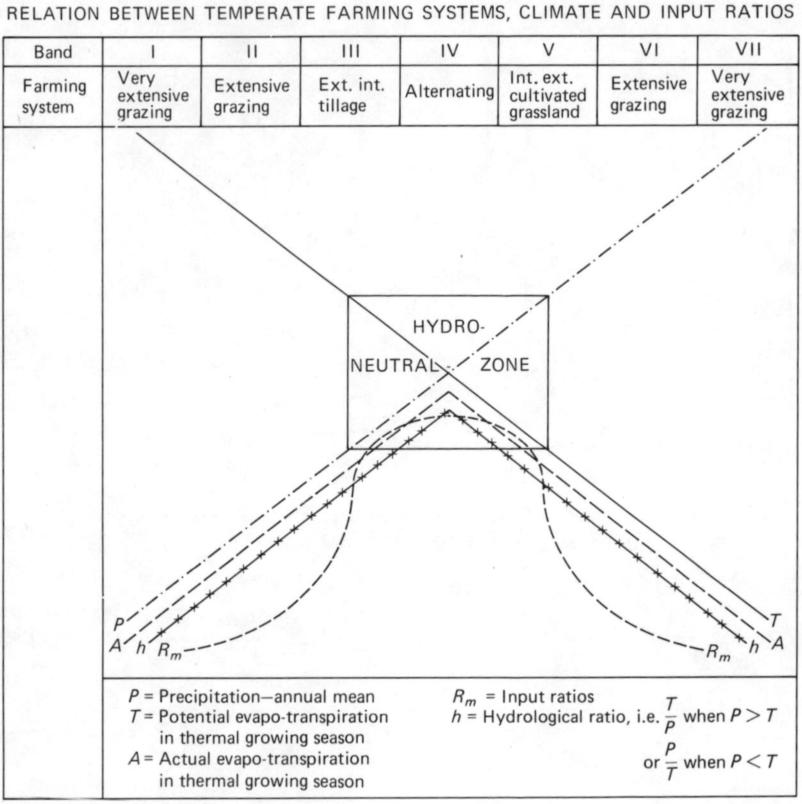

RELATION BETWEEN TEMPERATE FARMING SYSTEMS, CLIMATE AND INPUT RATIOS

Band	I	II	III	IV	V	VI	VII
Farming system	Very extensive grazing	Extensive grazing	Ext. int. tillage	Alternating	Int. ext. cultivated grassland	Extensive grazing	Very extensive grazing

HYDRO-NEUTRAL ZONE

P A h R_m R_m h T A

P = Precipitation—annual mean
T = Potential evapo-transpiration in thermal growing season
A = Actual evapo-transpiration in thermal growing season
R_m = Input ratios
h = Hydrological ratio, i.e. $\frac{T}{P}$ when $P > T$ or $\frac{P}{T}$ when $P < T$

FIGURE 8.3 Relation between temperate farming systems, climate and input ratios (from Duckham and Masefield, 1970).

and in cold, wet or dry areas; tillage, alternating and cultivated grassland systems, on the other hand, tend to occur in regions where either the mean annual precipitation (P) does not greatly exceed the mean potential evapo-transpiration (T) or where T does not greatly exceed P (see Fig. 8.3). These latter zones have been termed hydrologically neutral and represent regions within which there is substantial choice of farming systems.

Duckham and Masefield (1970) expressed these findings in a System Location Model, to predict the proportion of farm land devoted to tillage (S_t) and to grassland or grazing (S_g):

$$\frac{S_t}{S_g} = g\,.\,A.h - i(L_0 + L_m) + j\,.\,D_s - k\,.\,B$$

where g, i, j and k are constants and

$$h = \frac{T}{P} \text{ where } P > T$$

$$\text{or } \frac{P}{T} \text{ where } T > P;$$

L_o = the adverse effect of local difficulties (e.g. awkward relief, frost pockets, liability to flood, unworkable clay soils, etc.) and

L_m = the adverse effect of difficulties of access to market (including transport facilities, actual mileage, tariff, quota and exchange rate barriers, etc.)

D_s = Human population per unit of farmed land;

B = Index of socio-economic development, when U.S.A. = 100 on the Beckerman scale.

This model recognizes that many features of the human population and its activities play a part in the choice of farming systems, in addition to climatic conditions. For a number of reasons, it was not found possible to derive similar models relating to the choice of farming systems in the tropics.

Farming Systems in the Tropics

Ruthenberg (1971) has described the farming systems of the tropics and discussed the main ways in which systems of cultivation may be classified. Cultivation is here used in the sense of "the preparation and use of land for growing crops" and distinguished from "collecting". *Collecting* includes the regular and irregular harvesting of uncultivated plants. It may be carried out to add to the food gained from subsistence farming but only in a few cases is collecting a major cash-earning activity. Ruthenberg (1971) cites wild oil-palms in some parts of West Africa, gum arabic in the Sudan and wild honey in Tanzania, as examples of cash-earning, "collected" produce.

In addition to collecting and cultivation, Ruthenberg (1971) classifies livestock farming according to the degree of stationariness of both the animals and those who tend them.

The main categories are therefore:

(a) Total nomadism;
(b) Semi-nomadism;
(c) Transhumance;
(d) Partial nomadism;
(e) Stationary animal husbandry.

The choice of system from amongst these various possibilities has been related by MacArthur (1971) to the factors listed by Duckham and Masefield (1970) and discussed earlier in this chapter. MacArthur postulated the main paths along which system development might take place in the four major climatic zones of the tropics. These paths are shown in Fig. 8.4, in which the thickness of the arrows indicates their relative importance and feasibility.

The tendencies shown are accompanied by changes from:

(i) long-fallow to short-fallow systems to permanent land use;
(ii) low-intensity to high-intensity crops, and
(iii) arable farming to the planting of perennial crops, and the increasing use of cultivated fodder, irrigation, multiple cropping, fertilizers and tractors.

1. Tropical rainforest

Shifting cultivation
Semi-permanent cultivation
Permanent rain-fed farming
Irrigation farming (paddy rice)
Perennial crops

2. Humid savannas

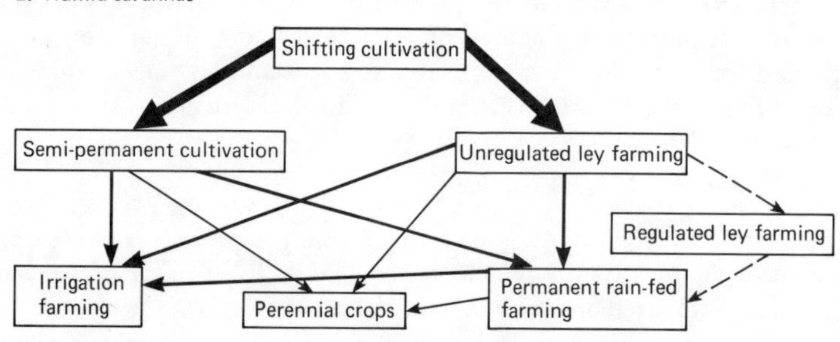

3. Dry savannas

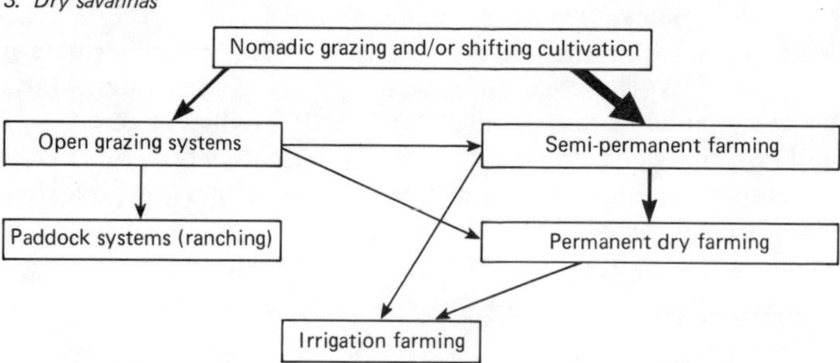

4. High altitude

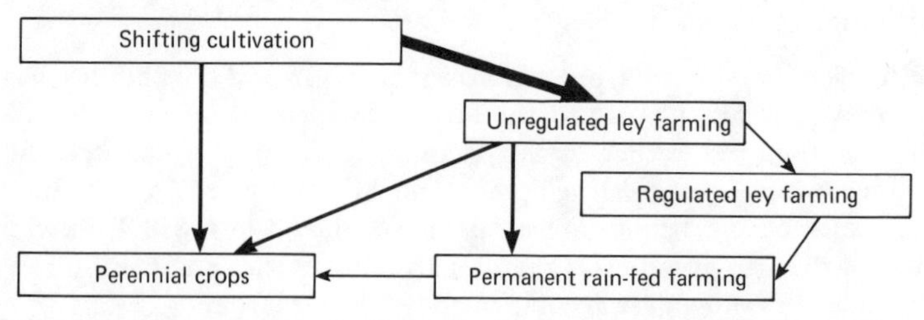

FIGURE 8.4 General tendencies in the development of tropical farming systems. These tendencies are valid for indigenous smallholders only. The establishment of large farms and estates is usually connected with the *ad hoc* adoption of an advanced system of farming (from MacArthur, 1971).

These changes arise from alterations in such important social and economic factors as population density, technical progress, the development of urban purchasing power and export markets, and human aspirations (MacArthur, 1971).

The Influence of the Farming System on the Environment

Farming systems may influence the environment in several ways. They may add or remove substances, change the appearance of the landscape and exert an influence on man by their manner of operation. In the main, however, the substances used by agriculture are not usually considered of great significance unless they change the appearance of the environment. This does not mean that they are unimportant but that they do not yet excite attention.

For example, the use of water in agricultural operations is not seen as a threat to water supplies for more important purposes. Land is regarded as properly used for agricultural purposes, rather than as an undesirable competition for land for other uses, such as building, forestry or even recreation. By contrast, some of the outputs of farming are very seriously criticized as pollutants of the environment (see Headley, 1972, for a discussion of the relations between agricultural productivity, technology and environmental quality).

Pollution

Definitions of pollution tend to be very wide* and thus to include almost any adverse change in the environment. This raises some difficulty in cases where there is no general agreement as to what is adverse and what is not.

Clearly different people think differently about such subjects as amenity values and scenic beauty, noise, crowds and smells. Even poisonous substances may only be so at certain concentrations and almost any substance is either harmful or undesirable if present in sufficient concentration. Thus, with a very wide definition, virtually anything can be a pollutant in certain circumstances. This suggests that such wide definitions are not helpful, since they cannot delineate anything in particular. Perhaps no definition of pollution can avoid the fact that conditions have to be stated, however. It is always necessary to specify the person or other living organism to whom something is a pollutant and it is necessary to state the circumstances under which this is so. In general, we have in mind man and those organisms he favours and the circumstances are those normally found where the pollution is said to occur.

Even so, a narrower definition might be more useful, such as: pollution is the

* "Environmental pollution is the unfavourable alteration of our surroundings, wholly or largely as a by-product of man's actions, through direct or indirect effects of changes in energy patterns, radiation levels, chemical and physical constitution, and abundance of organisms. These changes may affect man directly, or through his supplies of water and of agricultural and other biological products, his physical objects or possessions, or his opportunities for recreation and appreciation of nature".

— U.S. President's Science Advisory Committee, 1965.

presence of substances in concentrations or quantities that are injurious to living organisms. There are two important features of this definition. First, it recognizes that all living organisms may be subjected to environmental pollution but that *only* living organisms are so affected. A view may be spoiled by pylons but it is not, by this definition, a case of pollution. What, in fact, would be gained by so labelling it? Secondly, only substances can be pollutants; organisms cannot be. Terms such as pests, parasites and weeds already cover undesirable numbers of living organisms.

The clearest examples are those that involve substances that are poisonous to man himself or to the fauna or flora that he values, even in relatively small concentrations. These are not necessarily confined to the locality in which they are produced and, indeed, many of the most serious pollution problems are very wide-ranging and even world-wide. Pollutants that are present in air and water are particularly prone to wide dispersal.

Pollution in Air

The major air pollutants are of industrial origin. They include gaseous (e.g. SO_2), liquid (e.g. H_2SO_4 derived from SO_2) and solid (smoke, soot and dusts) constituents and may be carried for considerable distances and at considerable atmospheric heights.

Agricultural practices may give rise to local concentrations of methane and other gases but these are generally of nuisance value only. The most marked examples are probably offensive odours in the region of intensive animal enterprises, such as piggeries. Solid particles are more serious, however. Blowing dust, from wind erosion of soil, and from operations such as grass drying, may be of concern to the community; grass drying, incidentally, may also offend on grounds of excessive and unpleasant noise. The effects of air pollution on plants and on farm animals have recently been reviewed by Jones and Cowling (1974).

Many of the primary pollutants may give rise to secondary pollutants, resulting from interactions involving the primary substances.

Pollution in Water

This arises from agriculture in two main ways:

(a) by drainage from the land; and
(b) by drainage from animal houses or yards.

(a) Either by deep drainage or by surface run-off, substances applied to the land, or reaching the soil after being applied to animals or plants, may find their way into watercourses, lakes, reservoirs and, ultimately, the sea. This will apply to all soluble substances and some others carried away as fine particles. Quantitatively, the most important sources of water pollution from agriculture are fertilizers (Hook, 1972), animal excreta and chemicals applied as pesticides or herbicides.

TABLE 8.2 Quantities of excreta produced per day and the corresponding pollution loads from different animals
Volume and composition will vary considerably depending on type and weight of animal, feed, etc. (After Wheatland and Borne, 1970)

	Weight of animal (kg)	Reference	Quantity (l)	Dry matter (kg)	B.O.D. load (g)
Cow	500	MAFF (1968)	Urine 15		145
			Faeces 41·8		353
			Total 56·8	4·72	498
Calf	250	MAFF (1968)	Urine 7·5		147
			Faeces 20·9		200
			Total 28·4	2·72	347
Pig	68	–	Urine 1·7		–
			Faeces 2·8		–
			Total 4·5	0·5	137
Sheep	–	MAFF (1968)	Urine 1·8		55
			Faeces 3·0		47
			Total 4·8	0·545	102
Hen	2	Riley (1968)	Total 0·13	0·036	9

TABLE 8.3 Properties of animal manures (derived from Dodd, 1973)

Animal	Total N per day (g)	Total P per day (g)	Total K per day (g)
Fattening pig	29	7	9
100 laying hens	186	68	51
Fattening cattle:			
200 kg	54	6	49
450 kg	133	16	122
Cows	163	19	130

The most important constituents of fertilizers and animal excreta, in this context, are nitrates and phosphates: it is these that are responsible for the spectacular algal "blooms" that appear periodically, in lakes especially. Tables 8.2 and 8.3 illustrate the level of excretion of dry matter, nitrate and phosphate, that might be expected from the major livestock species, and the pollution load they represent.

Figure 8.5 illustrates the routes by which water pollution may arise from the use of fertilizers, but in order to put this question into perspective, it should be noted that the relative contribution of fertilizers to nitrate and phosphate pollution may be much less than contributions from other sources.

Problems due to algal growth in water as a result of nitrate and phosphate enrichment are of several kinds (see Hawkins, 1972). Algae may adversely affect

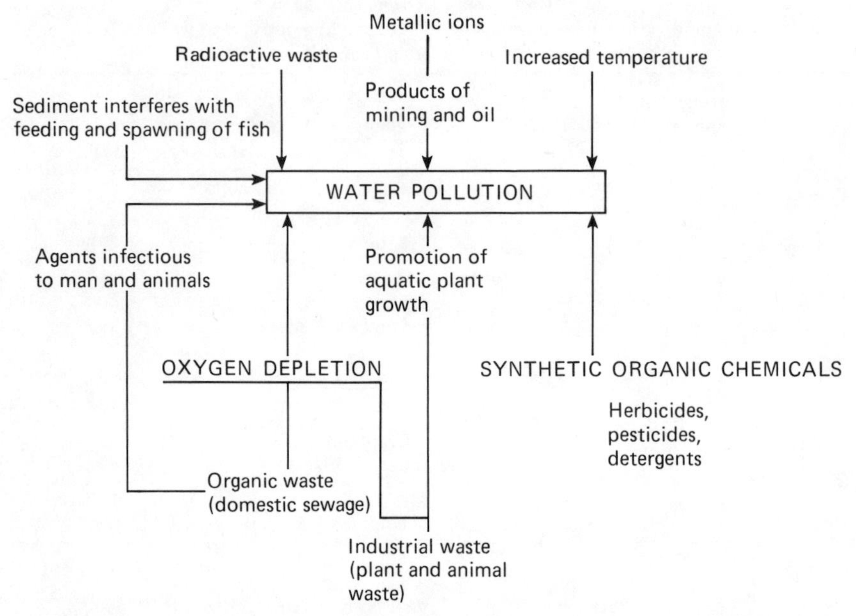

FIGURE 8.5 Interactions producing water pollution (from Mawdesley-Thomas and Fraser, 1972).

urban water supplies (by taint and odour, by clogging filters); kill fish with toxins and cause deoxygenation when massive algal populations die and decay; impede water movement in drainage and irrigation schemes; reduce the speed of ships by colonizing their hulls; and form serious weed problems in rice-paddies.

Eutrophication, due to nitrate and phosphate enrichment, is the major single cause of algal fouling problems; it is not confined to fresh waters but also affects coastal areas.

The major genera of nuisance algae have been listed by Hawkins (1972): they include green algae (Chlorophyta) such as *Chara* spp. in rice fields, diatoms such as *Asterionella* spp. that cause blooms in still water, other Chromophytes such as *Gonyaulax* spp. that are responsible for shell-fish poisoning, and blue-green algae (Cyanophyta) such as *Nostoc* spp. (a weed of rice) and *Anabaena* spp., another toxin producer.

Apart from algal "blooms", there is an additional hazard in the case of nitrate, due to contamination of drinking water supplies and the possibility of nitrate poisoning, especially of very young children (Watson, 1970). In most cases, pollution of the environment by agricultural practices involves the soil at some stage and much depends, therefore, on the capacity of soil to absorb or change the substances reaching it.

(b) Housed or yarded animals differ from those kept extensively, or even intensively-grazed, chiefly in their greater concentration per unit area. If their excreta is spread over the same area of land as they might otherwise have occupied, or over the area required to grow their food, there is little difference from the grazed situation, except perhaps in the pattern of return of excreta. It is usually more efficient, in terms of labour and equipment usage, to return the excreta at less frequent intervals and thus in greater quantities at any one time than would be the case in most grazing systems. It should never be supposed, however, that the distribution of excreta is anything like uniform under grazing conditions. Quite apart from the behaviour patterns of many animals in relation to faecal deposition (see Hafez, 1962), normal distribution does not mean that all areas receive some excreta. After 10 years of cow grazing, for example, Peterson *et al.* (1956) concluded that 6-7% of the pasture would have received no faeces, whilst about 15% would have received four excretions over the same period.

In many cases, of course, the concentration of housed or yarded animals is very high and it is not practicable to return their excreta to the land or, if it is, only to a rather small area. Pollution may arise in several ways, since the spread excreta itself, or the ditches and lagoons devised to treat it, may give offence to the eye and the nose, even when it is being efficiently deployed. Run-off effluent may reach waterways by different routes but causes pollution mainly because of its high biological demand for oxygen.

In terms of heavy metal pollution, agricultural slurry may contain considerable amounts of copper, but this is nothing like as objectionable as the quantities of, for example, lead or cadmium contained in sewage that may be returned to the land.

Clearly, no animal production system has been completely costed if the cost of detoxifying excreta has not been included.

Most agricultural pollution is via water and the chief pollutants are probably pesticides and nitrates. Although some 50% of the nitrogen applied to the land as fertilizer may remain unaccounted for, it does not necessarily reach waterways and is spread over a large area. It is the concentrations of housed and yarded livestock that pose the most spectacular (if localized) problems (Willrich and Smith, 1970). Some of the giant feedlots of America, for example, contain 10,000 cattle and produce vast quantities of waste. In some cases, the latter may be used to irrigate pastures (Albin, 1971) but many other ways of disposing of waste are being explored. Most are concerned with anaerobic or aerobic systems of decomposition in lagoons or oxidation ditches. Dehydration and burning, composting or some form of recycling are also being considered.

In addition to bio-degradation by earthworms, larvae of houseflies (*Musca domestica*) have been used on poultry manure, the insects being fed as a protein source to livestock (Miller and Shaw, 1969; Calvert *et al.*, 1970). However, dried cattle manure can have a direct use as part of the diet for poultry (Lipstein and Bornstein, 1971), or for feeding to cattle (Anthony and Nix, 1962; Anthony, 1970) and swine waste has been fed back to pigs (Anthony, 1971).

There is clearly no reason why good use should not be made of all these potentially polluting materials; it is largely a question of the cost of doing so.

Furthermore, agricultural systems may offer a means of absorbing pollution from other sources.

Pollution may be considered a special case of the effect of agricultural systems on human welfare. Clearly, any accumulation of poisons in the environment poses a direct threat to human well-being and must be seriously regarded. There are similar threats to the other living occupants of the environment, not only due to poisonous substances but also to the methods employed in agriculture.

In early agriculture, it was necessary to battle with other animals and plants that competed with man's interests. It is often argued that this is still the case, in terms of pests, parasites and weeds. The enormous scale of world pesticide usage has already been referred to (see Chapter 7) but it has been calculated (Chapman, 1973) that increasing farm acreage in the U.S.A. by 12% could cut pesticide use by 70-80%.

In the developed countries generally, man's control over the environment has now become so great that it is necessary to give special attention to the preservation of wildlife (Thompson, 1973).

Wildlife preservation is not only influenced by poisonous substances but also by the changes that farming brings about in the physical environment. Uncontrolled grazing may completely change the vegetation and thus the animals, both small and large, inhabiting the area.

In the U.K., the destruction of hedges in order to create large arable fields is bound to influence the numbers of plants, insects, birds and small mammals that normally inhabit them. It is sensible and not difficult, of course, to recognize this possibility but it is a quite different matter to assess the importance of such consequences. It seems entirely possible, however, to plan farms in such a way that wildlife is damaged very little (Barber, 1970; Powell, 1970) or even to combine wildlife preservation and farming with the addition of a game crop, such as pheasants and partridges (Ford, 1970).

Quite apart from specific effects on particular plants and animals, the general development of specialized agriculture is leaving behind a host of once-numerous breeds, now in danger of extinction. However, once the danger is established, there seems to be no difficulty in creating quite sufficient interest to ensure their survival, in farm parks and the like, provided that their numbers have not dropped too low (Henson, 1972).

Human attitudes to wildlife are extraordinarily mixed and, often, inconsistent. Some animals, including snakes, slugs and, notably, rats, are considered by most people to be pests with no good qualities at all: their destruction is regretted by few. Butterflies, on the other hand, are in a different category, generally regarded as adding value to life in spite of the fact that they do little material good and often do considerable harm. It is also true that most of us live in substantial ignorance of the existence of all but the most obvious plants and animals, even in our immediate environs. Yet it is possible to regret a reduction in the number of species and to worry when the numbers of any one species fall below some minimum level. It is curious that, in many minds, this reasonable concern for a species is visualized in terms of an image of a woolly, furry or feathered individual (that will certainly die and probably be eaten by another in the course of its individual battle for survival).

Wildlife are a significant part of the environment in which many people wish to live and ensuring their survival is one measure that can be taken to preserve or increase the amenity value of the agricultural landscape.

Amenity Use of Land

It is often argued that the demands of the community for land for amenity purposes will grow at an increasing rate and that agriculture will need to increase its efficiency of land use in order to produce even more from even less land.

This may well be true but concerns only the separate utilization of land for these two purposes. In addition, there are important ways in which

(a) agricultural land-use influences amenity values; and
(b) integrated use, for both agricultural and amenity purposes, is desirable.

There are also ways in which agricultural usage can contribute to amenity by converting unsightly land into attractive and useful areas. This applies to the waste heaps left in Britain, for example, as a result of mining for heavy metals, such as copper, zinc and lead. The reclamation of such land can be achieved by the use of organic residues or a layer of normal soil and eventual colonization by grasses. The process can probably be speeded up and cheapened by the use of tolerant populations of such grasses as *Agrostis tenuis, A. stolonifera* and *Festuca rubra* (Smith and Bradshaw, 1970).

a. Amenity Aspects of Agricultural Land Use

Since so much of the land is farmed in countries such as Britain, the appearance of agricultural land contributes substantially to the landscape and thus to the visual amenity of people living in or travelling through it.

Some landscape features of traditional farming, such as copses, shelter belts, and hedges, tend to disappear under intensive agricultural land use, particularly in large-scale arable farming. In Great Britain, the total length of hedgerows has been estimated at 616 000 miles or 1·1 mile per 100 acres (Locke, 1962) and they have been removed annually at the rate of about 0·73% of the total during 1946–70. Hedgerow removal is expensive and the biological effects of hedges are, on balance, considered to be favourable to the farmer, although this cannot easily be measured in economic terms (Caborn, 1971). Their influence on the appearance of the landscape is undoubted, however, especially in the U.K. Many other features have as much influence and some of these, such as irrigation reservoirs, may increase in number.

The planning of agricultural land increasingly takes account of wider landscape issues, even in countries like the Netherlands and Denmark, where the need for intensive use of land is as great as in the U.K. The incorpporation of woodland areas and shelter belts has also featured in recent plans for recasting the agricultural areas in Brittany and the Swabian Alps of Germany (Crowe, 1971).

A good case can be made in favour of development of the land for several purposes, in an integrated fashion, rather than the separate development of intensive agricultural use in one area and allowing land to go wild in another (Crowe, 1971); indeed, Prillevitz (1971) has argued that the density of human population, its mobility and the wider horizons of its members have made multiple land use of increasing importance.

b. Integrated Use of Land

There are many different ways of using land that might, in some circumstances, be sensibly integrated. In western Europe, an example may be found in multiple use of grassland. Grassland has a high recreational value: indeed, perhaps the best vegetation cover for land for most recreational purposes is grass-dominated herbage. Water also has many recreational uses and woodland has several: snow is best overlying grassland.

The grass sward is a self-renewing, self-cleaning cover, relatively dry and comfortable for sitting, walking, running and riding on. It is best kept short, however, and in many areas, including the U.K., would revert to scrub and woodland if it were not regularly defoliated. Recreational grassland requires management, therefore, including orderly defoliation. On small, smooth areas, that are not too wet or on too great a slope, cutting by mowing machines is possible and often preferable. On large areas, however, this is often not possible and rarely practicable or economic. These areas need grazing; but uncontrolled grazing may be too light or too heavy, and uncontrolled animal populations may increase too rapidly, resulting in overcrowding, undernutrition and disease.

A grazing animal population must in any case incur deaths, unless the numbers are methodically controlled, and corpses detract from the amenity value of the land, especially if individual animals are large (e.g. cattle). Since the grazing animals require management, including orderly culling, there is much to be said for integrated recreational and agricultural usage. Grazing animals in recreational areas have to be able to look after themselves to a very large extent: they must not demand a great deal of costly attention or supplementary feed. They have to be able to defend themselves against people or dogs, without being dangerous. They should also be relatively resistant to disease and to the kind of weather characteristic of the region. Cattle have many advantages where the grassland is good enough to support them; elsewhere sheep may be better. Sheep often suffer from disease, however, and require a great deal of attention, to their feet and at lambing and shearing times. They are also vulnerable to attack by dogs, although some benefit has sometimes been gained by including a proportion of goats.

At the Grassland Research Institute, in recent years, exploratory studies have been carried out on the Soay sheep (see plate I). These animals shed their coats naturally, so they require no shearing, and are very hardy and independent (Grubb and Jewell, 1966). They appear capable of surviving and reproducing without any attention at all and they can control grassland in a tussocky state. When crossed with a Down ram, they produce lambs quite unlike

PLATE I Soay ewes and their lambs grazing.

(*facing p. 198*)

PLATE II Soay ewes with their lambs (aged 8 weeks) by a Dorset Down ram.

(*facing p. 199*)

themselves (see Plate II), that are capable of good growth—even suckled on their dams—and produce acceptable carcasses. More work needs to be done before the usefulness of such an agricultural by-product from an "amenity" sheep can be assessed, but there are clearly interesting possibilities. It would not be necessary to leave the lambs with their ewes, if this were disadvantageous; they could form the basis of a quite separate agricultural enterprise.

Doubtless many other possibilities exist. The main point is that recreational land has to be managed and the skills of farmers could be employed with benefit to all. Indeed, it would be sensible for agriculturalists to be involved in the development of such ideas at the planning stage: it would be a pity if agriculture and other forms of land use were thought of as quite separate subjects.

The problems of land management in order to maintain a desirable flora occur in many parts of the world. In Africa, a great deal of use has traditionally been made of fire (Thomas, 1960), burning of the bush having been a major tool in the prevention or control of "bush-encroachment". In the high veld of south-central Africa, most grasslands are only temporary and the spread of trees and bushes has to be prevented if grazing animals are to be kept, whether by ranching of game or by the farming of domesticated animals.

It has been argued that biological control, using goats, would be the best way of dealing with the problem (Strong, 1973), involving greater efficiency and lower cost than other methods, depending on labour or the use of mechanical devices.

Similarly, in the *Pinus radiata* forests of New Zealand, grazing by farm livestock is practised as a part of forest management, before planting, during establishment and as the forest matures (Beveridge and Knowles, 1972).

Animal Welfare

There is one final interaction between the community and agriculture that should be considered. It concerns the increasing interest shown by the community in the way agricultural animals are kept (Bellerby, 1970; Taylor, 1972; Tudge, 1973; Blaxter, 1973). A highly emotional debate has grown up, mainly giving the impression that two sides exist who cannot see each other's point of view at all.

There are those who do not want animals to be kept for food, especially if it involves killing them: not all people who hold this view dislike consuming animal products. If animals are to be kept and killed, it would be generally accepted that the latter should be done humanely and that methods of animal production should not involve cruelty. But about what constitutes cruelty, there is much less agreement. Furthermore, it is at this point that people tend to lose contact with each other's arguments. Some take the view that modern intensive methods are undertaken to squeeze the last drop of profit out of the hapless animals involved. The operators of intensive animal enterprises naturally take a different view and consider their critics to be unrealistic, over-emotional and anthropomorphic.

The issue is often clouded by references to "unnatural" ways of confining animals, to which others object that agriculture (and wearing clothes) is unnatural and natural situations involve much suffering (whether you call it "cruelty" depends upon your point of view). One side tends to regard certain methods as obviously cruel and the other side gives the impression that since cruelty cannot be measured it can be ignored. Much debate hinges on the meaning to be attached to "stress" (Thorpe, 1971).

Such large differences of opinion are a very good indication of a shortage of facts but neither is there any general agreement about what facts are required. If we think about children, or dogs, however, it is clear that our judgements about cruelty and what is or is not tolerable do not depend upon objective measurements or the ability to make them. The fact is that we judge such situations in a highly subjective manner and individuals tend to draw the line in different places. But everyone draws the line somewhere and there may be said to be a "community" view about where it should be drawn.

In the case of farm animals, a similar situation exists, but it might help to recognize that everyone does draw the line line somewhere; it is largely a question of where. Furthermore, everyone decides subjectively, without measurement, but with more or less information and experience to help interpret what they see. Clearly, judgements are based primarily on observation, on what people see or think they see.

The great need, therefore, is for an improved capacity to relate observations to the notions of the kind of suffering that one wishes to avoid.

What freedoms or restraints should a child, or a dog, have? The difficulties of answering do not mean, however, that we can ignore the questions. So it is sensible to try and establish guidelines, or "Codes of Practice" (see Thorpe, 1969; Ray and Scott, 1973) for the keeping of livestock, to help decide where lines should be drawn between what is and what is not tolerable.

Agriculturalists should be able to help in this, in the same way as a child nurse ought to be able to help decide whether crying in a very young baby indicates suffering or not (we cannot ask *all* humans whether they are suffering). It should be possible to consider the ways in which animals are kept and to establish whether stocking densities, pen sizes or shapes, flooring materials, light intensities and many other such variables in the animals' environment adversely influence animal welfare, as indicated by behaviour (see Ewbank, 1969; Wood-Gush, 1973). For it is behaviour from which observation derives its information: the interpretation of such information is a legitimate, and not hopeless, area of enquiry.

It may be argued that animal-keepers would wish to pursue such enquiries anyway, but it seems likely that the rest of the community, or significant sections of it, will do so too. The most unfortunate outcome would be if agriculturalists and non-agriculturalists regarded each other as different in this context and did not talk to each other about it.

References

Albin, R. C. (1971). *J. Anim. Sci.* **32** (4), 803-810.

Anthony, W. B. (1970). *J. Anim. Sci.* **30** (2), 274-277.

Anthony, W. B. (1971). *J. Anim. Sci.* **32** (4), 799-802.
Anthony, W. B. and Nix, R. (1962). *J. Dairy Sci.* **XLV** (12), 1538-1539.
Barber, D. C. (1970). *World Wildlife News*, Summer, 1970, 22-23.
Bellerby, J. R. (1970). "Factory Farming". (J. R. Bellerby, ed.). Education Services, London.
Beveridge, A. E. and Knowles, R. L. (1972). *N.Z. Jl Agric.* **125** (1), 20-24.
Blaxter, K. L. (1973). *Vet. Rec.* **92**, 383-386.
Caborn, J. M. (1971). *Outlook on Agriculture* **6** (6), 279-284.
Calvert, C. C., Morgan, N. O. and Martin, R. D. (1970). *Proc. Maryland Nutr. Conf.* p. 69.
Chapman, D. (1973). *Environment* **15** (2), 12-17.
Clawson, W. J. (1971). *J. Anim. Sci.* **32** (4), 816-820.
Crowe, S. (1971). *Outlook on Agriculture* **6** (6), 291-296.
Dodd, V. A. (1973). "The Animal Manures Problem". Proc. CICRA Symp. Intensive Agriculture and the Environment.
Duckham, A. N. and Masefield, G. B. (1970). "Farming Systems of the World". Chatto and Windus, London.
Ewbank, R. (1969). *Outlook on Agriculture* **6** (1), 41-46.
Ford, C. (1970). *Farmers Weekly*, Oct. 23, 1970, ii.
Fosgate, O. T. and Babb, M. R. (1972). *J. Dairy Sci.* **55** (6), 870-872.
Gaisford, M. (1972). *Farmers Weekly*, Feb. 25, 1972, p. xii.
Gerard, B. M. (1967). *J. Anim. Ecol.* **36**, 235-252.
Grubb, P. and Jewell, P. A. (1966). *Symp. zool. Soc. Lond.* No. 18, 179-210.
Hafez, E. S. E. (1962). "The Behaviour of Domestic Animals". Bailliere, Tindall and Cox, London.
Hawkins, A. F. (1972). *Outlook on Agriculture* **7** (1), 21-26.
Headley, J. C. (1972). *Am. J. agric. Econ.* **54** (5), 749-756.
Heath, G. W. (1962). *J. Br. Grassl. Soc.* **17** (4), 237-244.
Henson, J. (1972). *World Wildlife News*, Summer 1972, 6-8.
Hook, S. (1972). "The Contribution of Fertilisers to Water Pollution". Unpublished dissertation, University of Reading.
Jones, L. H. P. and Cowling, D. W. (1974). *In* "Control of Industrial Pollution" (A. Parker, ed.), Vol. I "Air Pollution". McGraw-Hill, London (in press).
Lakhani, K. H. and Satchell, J. E. (1970). *J. Anim. Ecol.* **39**, 473-492.
Lipstein, B. and Bornstein, S. (1971). *Israel J. agric. Res.* **21** (4), 163-171.
Locke, G. M. (1962). *Q. Jl For.* **56**, 137.
MacArthur, J. D. (1971), *In* "Farming Systems in the Tropics" (H. Ruthenberg, ed.), Chapter 10. Oxford University Press.
MAFF (1968). Poultry Waste Conferences, Sunningdale.
Mawdesley-Thomas, L. E. and Fraser, W. D. (1972). *Br. vet. J.* **128**, 337.
Miller, B. F. and Shaw, J. H. (1969). *Poultry Sci.* **48** (5), 1844.
Murphy, L. S., Wallingford, G. W. and Powers, W. L. (1973). *J. Dairy Sci.* **56**, 1367-1374.
Perfect, J. (1972). *Biologist* **19** (3), 149-151.
Peterson, R. G., Lucas, H. L. and Woodhouse, W. W. (1956). *Pl. Agron. J.* **48**, 440-444.

Powell, L. B. (1970). *Farm and Country* Sept., 53-54.

Prillevitz, F. C. (1971). *Outlook on Agriculture* **6** (6), 285-290.

Ray, P. M. and Scott, W. N. (1973). *Br. vet. J.* **129**, 194-201.

Riley, C. J. (1968). *Water Pollution Control* **67** (6), 627.

Ruthenberg, H. (ed.). (1971). "Farming Systems in the Tropics". Oxford University Press.

Smith, R. A. H. and Bradshaw, A. D. (1970). *Nature, Lond.* **227** (5256), 376-377.

Spedding, C. R. W. (1972). *Proc. 2nd Wld Congr. Anim. Feed. Madrid.* **4**, 1245-1258.

Stockdill, S. M. J. (1966). *Proc. N.Z. ecol. Soc.* **13**, 68-75.

Stockdill, S. M. J. and Cossens, G. G. (1967). *Proc. N.Z. Grassl. Assoc.* 1966, 168-183.

Strong, R. M. (1973). *Biological Conservation* **5** (2), 96-104.

Taylor, G. B. (1972). *Vet. Rec.* **91**, 426-428.

Thomas, A. S. (1960). *Prox. IXth int. Grassl. Congr., Reading*, 405-7.

Thompson, H. V. (1973). *Br. vet. J.* **129**, 202-206.

Thorpe, W. H. (1969). *Nature, Lond.* **224** (5214), 18-20.

Thorpe, W. H. (1971). *New Scientist and Science Journal* Mar. 11, 1971, 560-561.

Tudge, C. (1973). *Farmers Weekly* Sept. 14, 1973, 93.

Watson, Sir S. (1970). *Advmt Sci.* **27**, 1-13.

Wheatland, A. B. and Borne, B. J. (1970). *Water Pollution Control* **69**, (2), 195-205.

Whyte, R. O. (1964). "The Grassland and Fodder Resources of India". Indian Council of Agricultural Research, New Delhi.

Willrich, T. L. and Smith, G. E. (1970). "Agricultural Practices and Water Quality". Iowa State University Press.

Wood-Gush, D. G. M. (1973). *Br. vet. J.* **129**, 167-174.

9 Agricultural Biology and the Community

The contributions that any subject may make to the community as a whole are many and varied and cannot be summarized briefly. The impact that they actually make, however, depends greatly on the way that they are studied and developed. Well-known and established subjects often go through phases when they appear inward-looking and, to the outsider, rather dull and arid. At other times, the same subject may appear, even to those who know nothing about it, to be exciting, full of interest and in a state of rapid change.

This public image of a subject may be thought of as an indicator of the impact that the subject is making on the "mind" of the community. Such an influence can be very important and affect, for example, the number of school-leavers who choose to study the subject and the degree of public support it receives, including money for its study and development. The image also affects the extent to which the subject and its exponents are "reputable" and thus the extent to which they are able to influence society.

There are, of course, many other ways in which a subject may influence the community. Some of them may result in the community changing its habits, its way of life or even its standards, without even being aware that it is being influenced.

Subjects such as agriculture, that are taught for both educational and vocational reasons, may clearly influence those educated who are not going to be engaged within the agricultural industry, as well as those whose careers fall within one or another branch of agricultural practice.

Since the practice of agriculture has many effects on the non-agricultural community and since all people depend primarily upon agriculture for their food, it would be advantageous if knowledge of agriculture were not confined to those who practise it. Indeed, it can be argued that no educated person should be ignorant of the way in which civilization depends upon a highly productive agricultural base. Tests of literacy and numeracy have been debated for a long time, but it does not seem to have been generally appreciated that the role of agriculture is so central to society that ignorance of it might be said to indicate an incomplete education.

In case this seems to be an exaggerated view, it is worth considering the remarkable symbiotic relationships that man has built up with the species of agricultural plants and animals.

Farming is not a highly-controlled industry in the main; it still depends, for the most part, on working with natural forces to exploit living processes to our own productive ends. If the basic resources should be sufficiently misused by an ignorant, thoughtless or careless community, our capacity to feed ourselves might prove totally inadequate and there is no current alternative to agriculture for supporting the rest of society's activities. It is possible that it will one day be otherwise but, at the present time, anyone who really believes that milk comes out of a bottle or who has never considered what cows live on in the winter may be counted as uneducated.

In underdeveloped countries, there is a great need to bring new knowledge, already available, to peasant farmers: indeed, this has been described as "probably the most productive investment which can be made in any of the poorer agricultural economies" (Coverdale, 1972). One way of helping this process would be to teach biology in rural schools in such a way that its relevance to local agriculture could be clearly seen. This could usefully include practical projects involving the growing of crops and the keeping of animals. In this context, the usefulness of the subject is obvious but its educational contributions are not limited to situations of this kind.

In fact, agriculture as a subject has many additional educational qualities. It is inevitably multidisciplinary, it mixes science and non-science, it involves people, it involves economics and it should concern everybody, if only because it is a major land user.

These features may not be accepted by everyone as positive advantages, however; the main reasons for including them are as follows. In order to study parts of our environment it is necessary to establish separate subjects or disciplines, simply because of the difficulties of trying to cope with everything all the time. But the need for such separation should never obscure the fact that it is an artificial device to be employed where it is useful and to be dropped where it is not.

There are considerable educational advantages in a subject where the need for both specialists and integrators is obvious. Specialists are required to provide detailed information and a deep understanding of components. Integrators are sometimes called "generalists" but there are often overtones of superficiality associated with the term, even though general propositions have achieved the height of scientific respectability as "Laws".

Dobzhansky (1966) made a distinction between "reductionists" and "compositionists" and pointed out that they are complementary. He gave two splendid illustrations of the latter. The first concerned animal physiology: "to understand respiration in vertebrates, a knowledge of the structure of haemoglobins is indispensable; a knowledge of the physiological and ecological functions of respiration is indispensable to understand the haemoglobins". The second simply pointed out the difficulty of deducing the properties of alcohol from a knowledge of hydrogen and oxygen, both of which are normally gases, and carbon, which is solid and black in charcoal but solid and light-coloured in diamond.

The study of a grazing ruminant, for example, cannot be accomplished by a botanist or a zoologist or a biochemist alone. The animal and its pasture interact with each other and, even if they did not, the ruminant animal, by definition, has a stomach full of tiny plants that enable it to live on pasture.

It is all very well to argue that teams are needed—and so they are—but teams need leaders and the members have to be able to communicate with each other. There are many such areas of great importance that involve more than one discipline. The value of a multidisciplinary subject is that it not only deals with some of these areas but it demonstrates that they exist and that they can be studied.

The value of mixing science and non-science, and of mixing the study of people and of economics, is that educational systems sometimes tend to produce people who have only been exposed to one or other of these pair-members. This "bridging of the gap" is best performed naturally by a subject like agriculture, which genuinely involves them all.

Finally, every citizen should be concerned about the way the land is used,

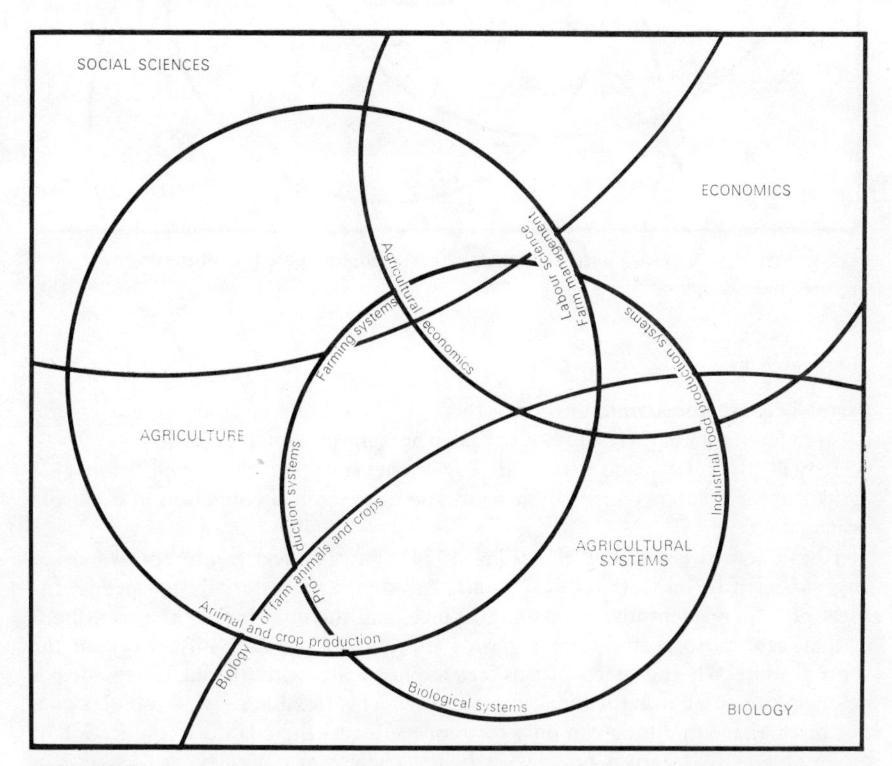

FIGURE 9.1 Agricultural systems as a subject and its relationship to other subjects: based on the idea that subjects are not mutually exclusive but represent overlapping areas of interest, each surrounding a unique focal point (or centre); and having regard to the fact that such relationships cannot be wholly satisfactorily illustrated in two dimensions, or in a simple diagram at all (from Spedding, 1971).

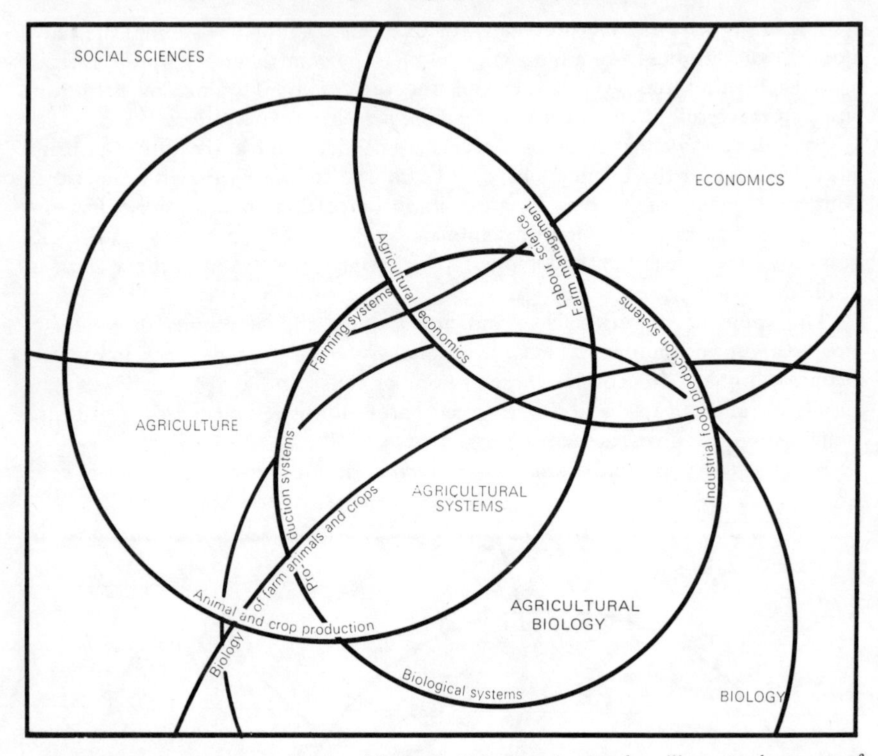

FIGURE 9.2 Fig. 9.1 with an additional circle superimposed to illustrate the scope of Agricultural Biology.

partly because the community is affected by methods of land use but chiefly because land is a major feature of the environment in which we live.

It will be recognized that this argument rests on the proposition that agriculture is a subject in its own right, and not merely a collection of bits from other subjects.

I have argued elsewhere (Spedding, 1971) that a subject can be formed about any sufficiently important focal point. All that is necessary is to specify the focal point, and demonstrate its importance, and then to identify the area about it that can most usefully be regarded as exerting relevant influences on the central point. Whether parts of this area are normally or traditionally regarded as belonging to some other subject is immaterial. This flexible view of subjects does not insist that something can only be in one subject: there is thus no question of quarrelling as to where it belongs—it belongs wherever it is relevant. If screws or nuts were only ever considered in company with other screws or nuts (the appropriate subject groupings), it would never be possible to consider microscopes or watches.

In thinking about agricultural systems, the same argument has been applied, and the overlapping nature of subjects has been illustrated in Fig. 9.1. In Fig. 9.2 a further circle has been added to super-impose, so far as this is possible, the subject of agricultural biology.

Agricultural Biology as a Subject

In its widest sense, agricultural biology must include the biology of all the agricultural animals and plants and of those non-agricultural organisms that are important in agriculture, as pests, parasites or weeds.

But many of the latter, particularly, and some of the former, would be included in other branches of biology: pest, parasite and weed species, for example, exist in their own right and are of interest quite apart from any agricultural connotations. The details of their biology really belong elsewhere, therefore, and it is their impact on agricultural systems, and the impact of these systems on them, that is the main concern of agricultural biology.

So the emphasis in this subject is on the way in which the relevant biological systems fit in to agricultural systems, how they operate, what effects they have, what affects them and in what ways, how they can be controlled or manipulated, their role and importance in agriculture. To understand these relationships it may often be necessary to know a good deal about the biology of the component organisms and, as mentioned earlier, agricultural biology in its widest sense would include all this. That is why this book is not entitled "Agricultural Biology", since it is restricted to a discussion of the role of biological systems in agriculture and how they function as parts (or sub-systems) of agricultural systems. The main argument is that their role can only be understood in relation to the agricultural systems of which they are a part and that it is necessary to understand sufficient of these systems—their purposes, products and processes—in order to do this.

Agricultural biology thus includes both the material and the viewpoint presented here but may also include much detailed biology of the constituents. However, it is not merely the addition of detail that distinguishes the larger subject, but the need to deal systematically and comprehensively with the biological systems of agriculture.

In this book, biological systems have been selected to illustrate particular points but there are many important ones that have not been mentioned at all and there are many that have only been named and not described.

The essentials of the subject are perhaps more easily seen in this way and some picture of the essentials seems desirable as a framework on which to hang more detailed information acquired later. It does not follow that this is the best or the only way in which the subject can be approached, studied or taught. Indeed, anyone starting from any particular point may already be a well-informed zoologist, botanist, parasitologist, soil chemist, rumen physiologist or plant biochemist. The one common need is for a coherent picture of the whole systems to which their specialism is intended to contribute.

However, if the subject of agricultural biology is considered in the same way

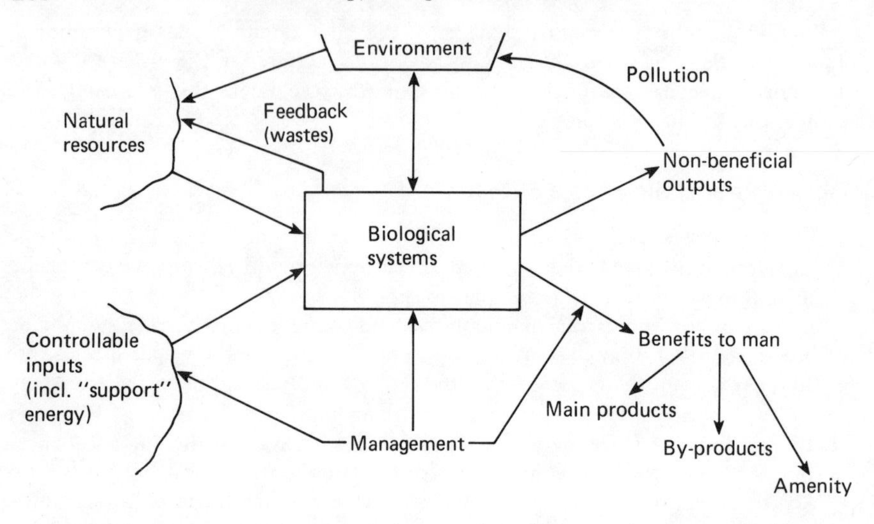

FIGURE 9.3 The role of a biological system in agriculture. The diagram could apply equally well to forestry or brewing, however, and is not therefore uniquely agricultural.

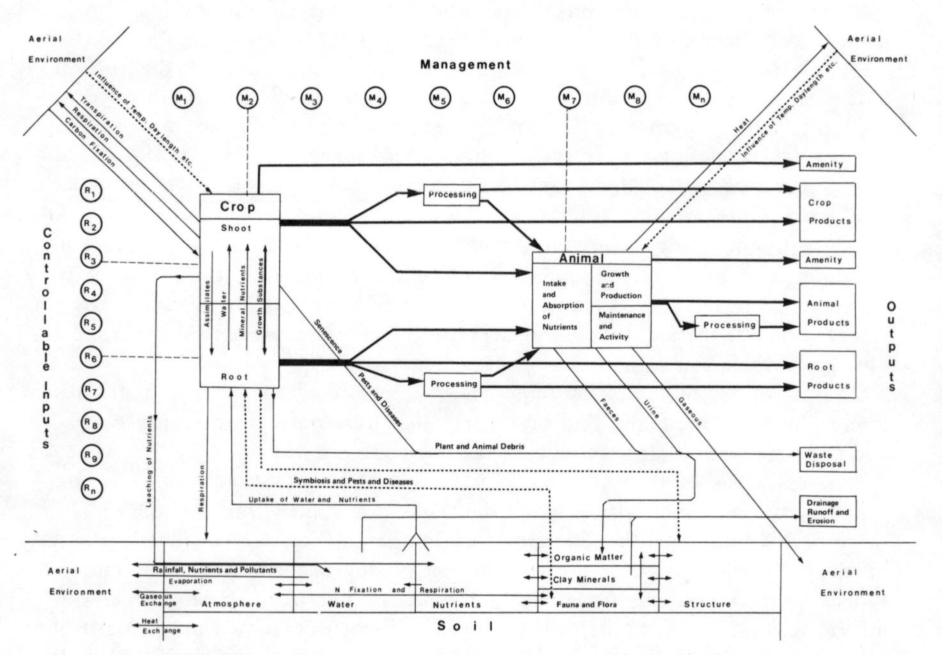

FIGURE 9.4 An elaboration of Fig. 9.3 to show greater detail, but omitting most of the relationships between management variables and the biological systems portrayed, because they depend upon precisely which system is envisaged.

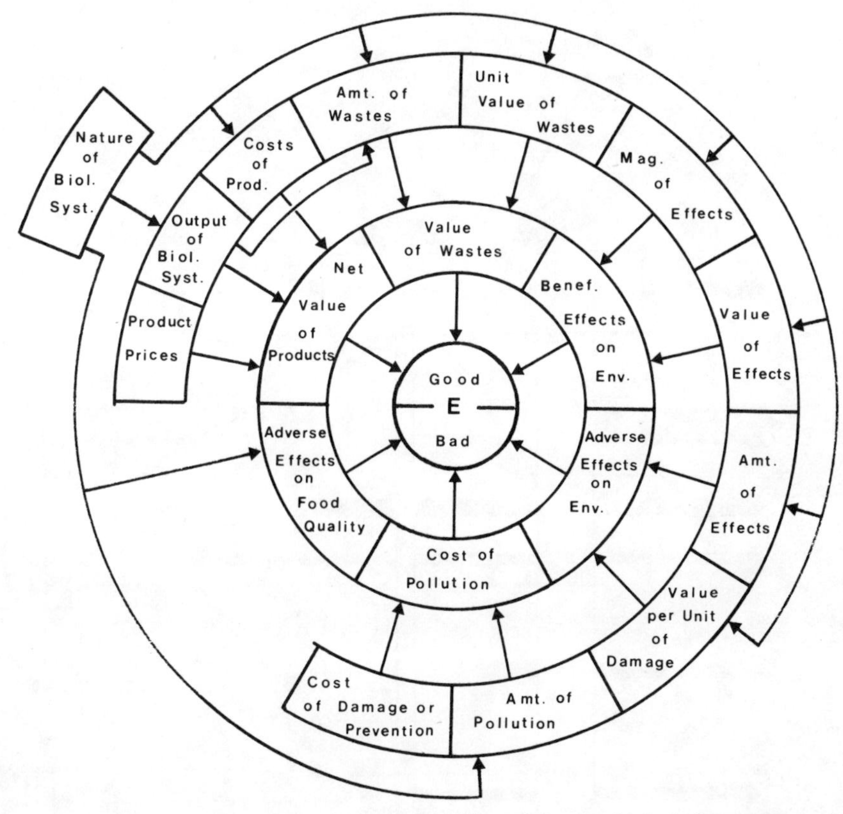

FIGURE 9.5 A diagram to include the effects (E) of a biological system on the community.

KEY:

E.	Effects	Amt.	Amount
Biol. Syst.	Biological system	Mag.	Magnitude
Prod.	Production	Benef.	Beneficial
Env.	Environment		

as, for example, agriculture, to be taught from the beginning (for instance, as a University degree subject), then it might be better to aim initially at a grounding in the contributory disciplines. It is then highly desirable that this groundwork should be oriented towards the edifice for which it is a foundation and should not contain masses of irrelevant material. The building analogy is quite a helpful one. However important the foundations and however obviously they have to come first physically, they clearly need to be preceded by plans and pictures of the whole building. Without a picture, no-one knows what it is going to look like and without detailed plans it is impossible for the right foundations to be laid. It is quite possible for foundations to be excessive as well as inadequate and their construction tends to be uninteresting if no picture of the final outcome is provided.

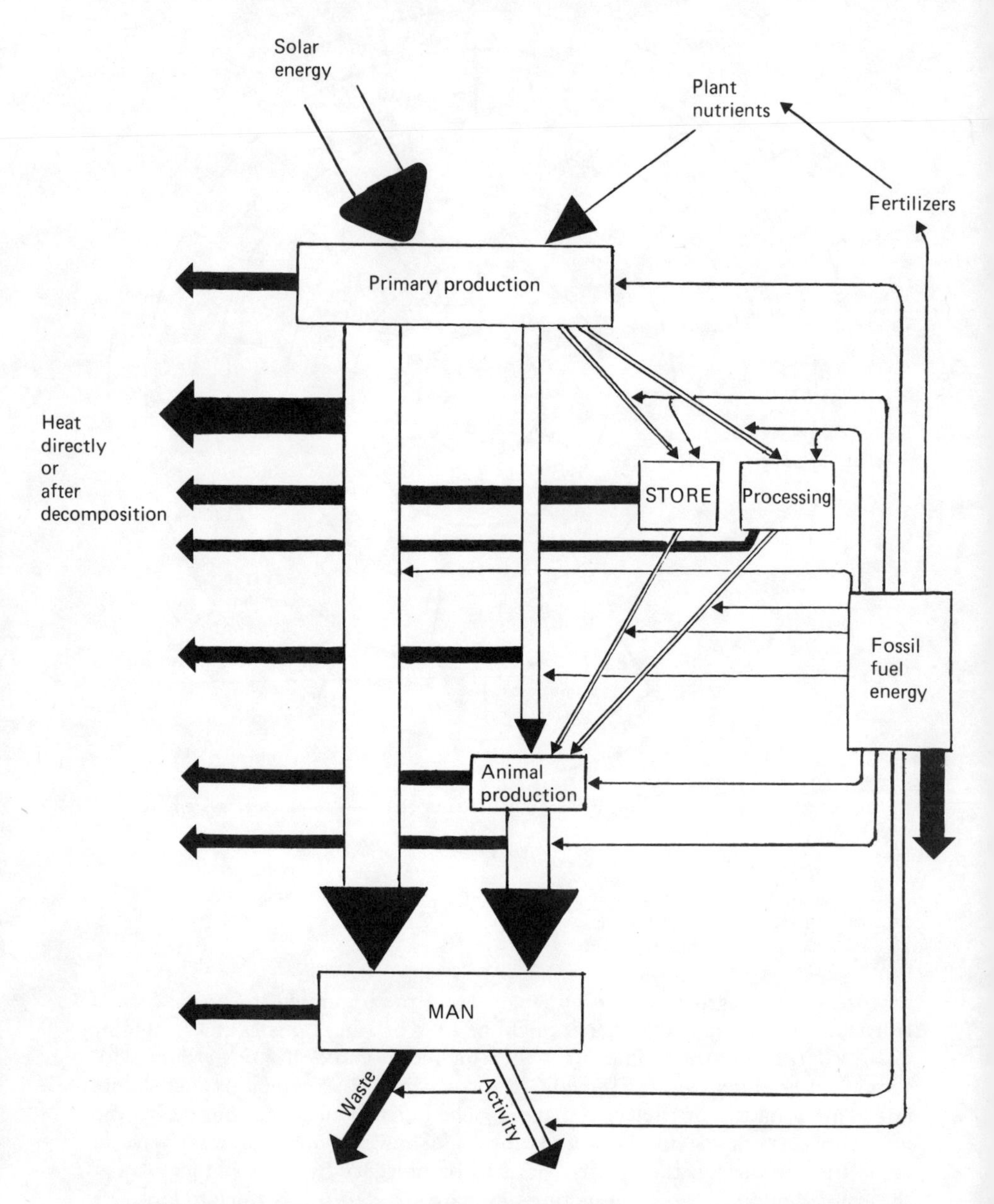

FIGURE 9.6 Energy flow in intensive agriculture. All processes produce waste (shown as shaded arrows), directly as heat or leading to decomposition and most processes involve inputs of fossil fuel energy (in less intensive systems, this energy is provided by man and animal power).

Unfortunately, as was pointed out in Chapter 2, it is not a simple matter to provide such a picture and, indeed, *one* picture is unlikely to be sufficient. It is easy to recommend such ideas without ever facing up to what they actually mean in practice: for this reason, an attempt has been made in Figs 9.3-9.7 to

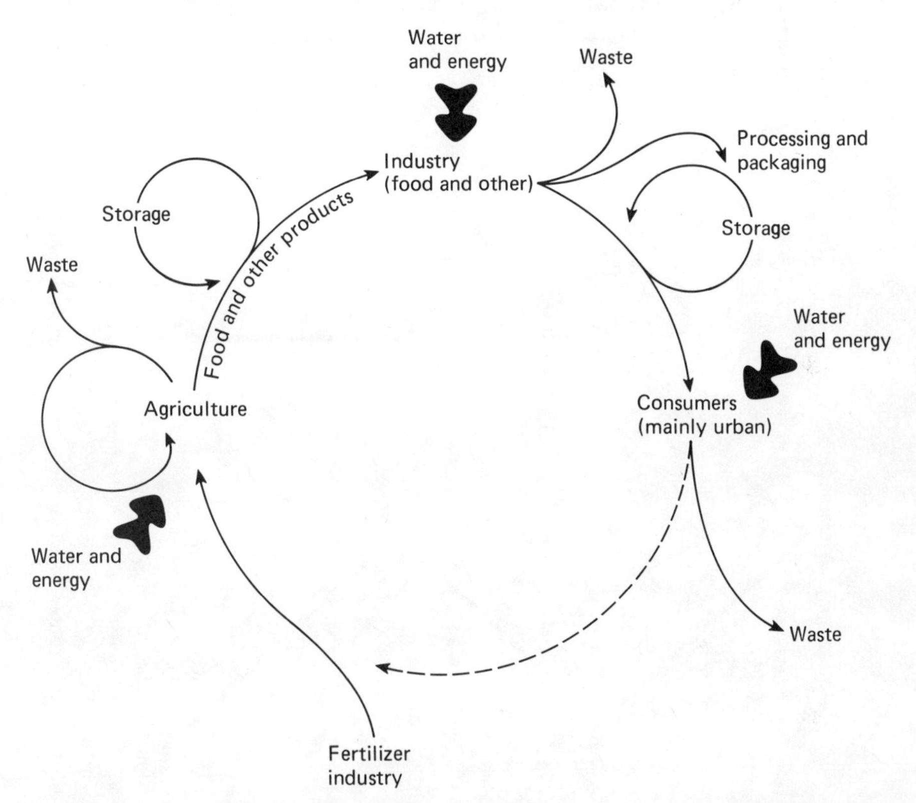

FIGURE 9.7 The major cycle of food production and consumption in industrialized communities, showing only small returns of waste to agriculture (see also Borgstrom, 1973).

illustrate what agriculture is all about. It has been assumed that the first picture has to try to be complete and must therefore be as simple as possible. Further elaboration is either limited to parts of this first picture or can be related back to it for clarification. These diagrams relate to the present functions of agriculture. Figure 9.8 attempts to illustrate the way in which agricultural activities developed from food collection and hunting.

All this is not difficult to relate to the needs of those engaged within agriculture but it may be less obvious why such a subject should be of interest to others.

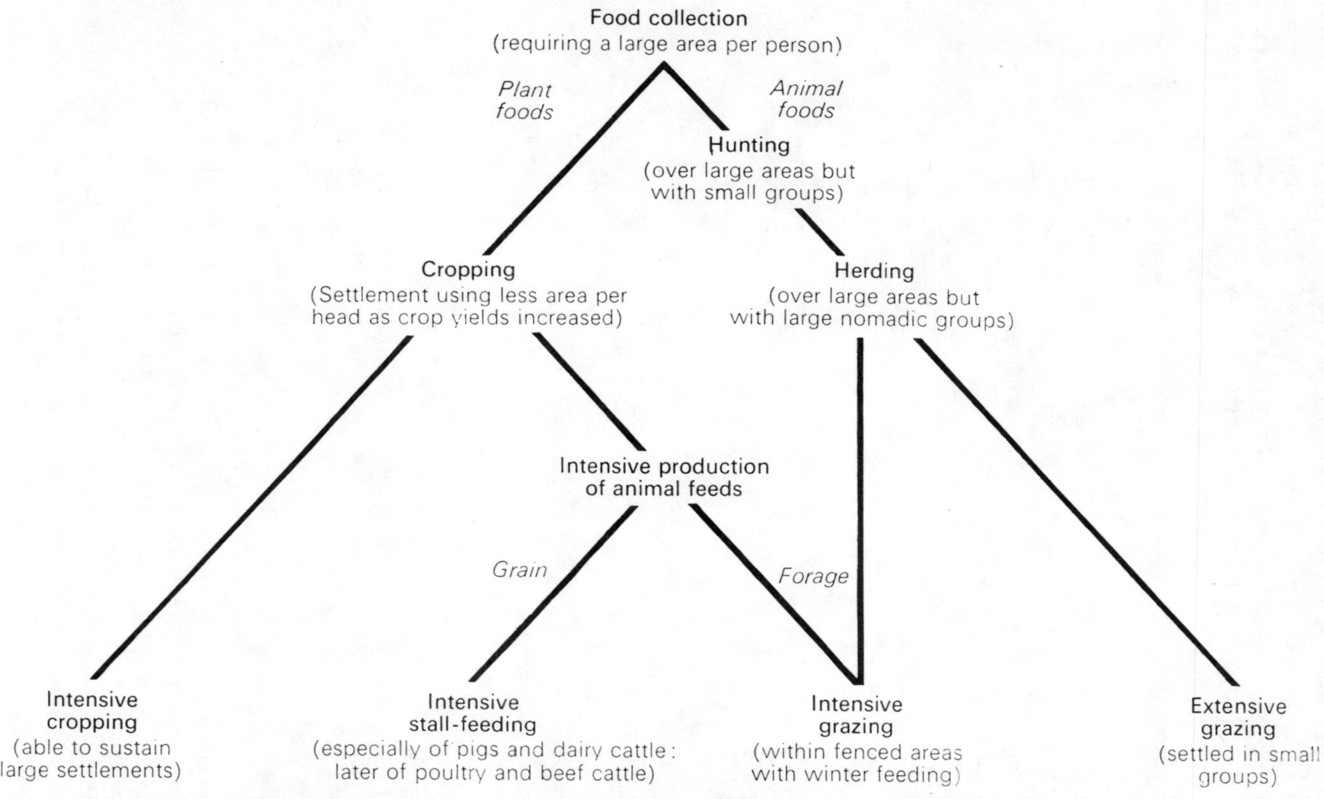

FIGURE 9.8 Development of agricultural systems: a simple diagram to illustrate major developments (from Spedding, 1971).

The Relevance of Agricultural Biology

Several reasons have so far been suggested as to why non-agriculturalists should be interested in agriculture, especially as part of the whole question of resource management, where the major resources used are components of everyone's environment, such as land and animals. The use of the environment affects us all in many ways, not least because its major uses tend to influence the shape of the environment itself and it is, or should be, a community concern to ensure that this shaping is in desirable directions. But the community operates, in planning and organizational matters, through its educated individuals, and their capacity to influence matters of this kind depends upon their knowledge and understanding of the issues involved. Understanding is required, rather than simple information, because influencing the environment usually incurs a cost.

The community may develop an increasing interest in the ways in which farmers use their land and may hope to influence them in the direction of conservation of amenity values, from scenic views to wildlife. Where people of different backgrounds have joined together to plan a farm with all these interests in mind, one major difficulty has been to work out how any additional costs might be met (see *Farming and Wildlife in Dorset*, 1970). In all such planning, it is not merely a question of how we wish to plan and shape our environment but also of how much we are prepared to pay for a given solution.

This is also relevant to another developing area of interest, our relations with the other animals inhabiting the planet. It will have been clear from the discussion in Chapter 8 that a civilized community is ultimately going to be concerned with the way in which agricultural animals are kept and killed. This is simply part of the nature of civilized peoples, to examine the bases upon which their culture is founded. It is part of the present argument that society should be aware of the sources of its wealth and prosperity and should understand the nature of the agricultural processes from which it derives most of its food. Such an understanding involves more than the biology of these agricultural processes, but this is the foundation subject upon which the rest must be built.

There is also a sense in which the study of agricultural animals may help in sorting out our relationships with the non-agricultural fauna. In agriculture, the reasons for keeping animals are quite clear and most people would accept that we exploit them for our own purposes; the same is true of crop plants but the overtones of exploitation are different. Even the exploitation of animals looks a little different in the light of natural history, however, since animals naturally eat each other without respect for age, sex and frailty and without regard for suffering or bereavement. It is hard to sustain the view that there is anything wrong or undesirable in eating our fellow animals, and keeping them healthy and comfortable before doing so can hardly make it worse.

Methods of keeping animals, once we have accepted responsibility for looking after them, are rightly a matter for concern, as is the method of slaughter. Equally, the study of agricultural animals demonstrates quite clearly that populations do not automatically and simply maintain the right balance of sexes and the numbers of individuals appropriate to their food supply without considerable mortality and stress. Conservation means controlling the environment in such a way that these natural forces can operate with the minimum of human intervention. We thus condone the natural order and, indeed, for very small creatures, we really have no choice.

The fact that we have to recognize exactly what is happening and what is necessary in agriculture should increase our insight into the natural ecosystems within which we live.

References

Borgstrom, G. (1973). *Ambio (Sweden)* **11** (5), 129-135.

Coverdale, G. M. (1972). *J. biol. Educ.* **7**, 40-46.

Dobzhansky, T. (1966). *Am. Nat.* **100** (915), 541-550.

Farming and Wildlife in Dorset. (1970). Report of a study conference, The Dorset Naturalists' Trust, Poole.

Spedding, C. R. W. (1971). *Outlook on Agriculture* **6** (6), 242-247.

10 The Role of Research

Subjects are not, or should not be, static and the biology of agricultural systems is certainly in a rapidly developing phase.

Research is an activity designed towards further development, but not merely by an extension at the boundaries of existing knowledge. The idea that there is a central core of knowledge that can be regarded as known for certain for a given subject, that there are fluid areas that have not yet crystallized out and other areas of great ignorance, is an unhelpful over-simplification. It suggests that research is confined to the fluid areas and not concerned with the central core; furthermore it suggests that it can be considered separately. The reason for this chapter is to argue that research is an extricable part of a developing subject and should not be so separated.

Research as an Integral Part of the Subject

The fact is that subjects do not simply have central cores of certainty and peripheral uncertainties. The centre itself is in a state of continuous change, but in any case is not a clearly worked-out, unambiguous, well-organized body of knowledge. The nature of a subject is really much more hazy and the subject content consists of facts and relationships, some of them hypothetical.*

The result is that research and re-thinking permeate all parts of a subject and the centre is as much the subject of continual reclarification and restatement as are any of the more peripheral areas.

It follows quite naturally from this that there cannot be a sharp separation between the subject as taught and that part of it that is the concern of research workers. Any such separation leads to an over-dogmatic view of teaching, the material of which becomes arid and lifeless, and it also tends to lead to a restriction of research activity to the less important parts of the subject. If this kind of situation develops, it is bound to be followed at some time by a major upheaval, during which the subject appears to undergo a violent change brought about by a revolution external to it.

* The nature of scientific knowledge and how it is increased represent a very large area of discussion; the following references will serve as a useful point of entry: Popper (1959); Kuhn (1962); Waddington (1968-70); Koestler and Smythies (1969); Lakatos and Musgrave (1970); Ravetz (1971).

It is this essential integrity of a subject that is the basis of arguments that teaching, research and advice (or extension) should not be separated. Agriculture has been treated in a variety of different ways in different parts of the world and it is inevitable that some people should specialize in one branch or another. The problem is largely an organizational one of maintaining contact between them but the importance of this will only be clear if the essential integrity of a subject is recognized.

Applied research is a good example of an area in which this is not always recognized and, of course, it is always possible to conduct some useful research of a trial and error nature that has little to do with the development of subjects: it is more concerned with the development of practical expertise. However, the ability of research workers to solve practical problems depends ultimately on two quite different contributions. The first is an adequate understanding of all the processes involved in the practical situation, at an appropriate level of detail. The second is a clear statement of the problem in terms that make it susceptible to research. Each of these contributions might seem to come naturally from different kinds of people and yet each requires some familiarity and understanding of the other area. Some of the implications of these propositions will be discussed in a later section; the relevant point here is that all activities within a subject are linked and there are risks in cutting them off from each other. If they are cut off, applied research becomes superficial and more basic research may become totally irrelevant to practical needs.

Another aspect of this relates to the contributions that one subject may make to another. For example, agriculture is sometimes thought of as a practical subject to which sciences such as botany, zoology and biochemistry may make important contributions. However, for the reasons mentioned earlier, research in these subjects will be primarily concerned with the development of those same subjects: it will be largely a chance bonus if results of such research have direct importance for agriculture. The main research that will benefit agriculture is *agricultural* research and this requires that the scientific basis of agriculture be recognized as a subject. Incidentally, it is not implied here that agricultural research is not done, or *should* be done, by, for example, zoologists and botanists.

As discussed in Chapters 2 and 3, agriculture and agricultural systems are more than just scientific subjects; they involve people, management, economics, processing and marketing. But they have a scientific basis, largely biological, that is different from botany, zoology and all the other disciplines involved. It is this subject that needs to be developed and within which research is needed.

It is not suggested that it does not exist anywhere or that no such research is being carried out, but this way of looking at it is certainly not universally accepted or even consciously recognized even by those who tend to operate in the manner advocated. It *is* suggested that conscious recognition would allow the subject to advance and develop more rapidly and would encourage the necessary integration of different contributory specialisms.

This is a restatement of the proposition put forward in Chapter 2 that a subject could usefully be defined simply as a relevant area surrounding a focal point of major concern. The argument now is that research in a subject so defined is an integral part of it. Thus the biology of agricultural systems has a research content that, like that of any other practical subject, needs a continuous range of activity from the most applied to the most fundamental.

Many people would draw the line at the latter but scientific curiosity can be allowed free play within any subject: it is a contradiction in terms to suppose that it can be confined to traditional disciplines. Others suppose that, whilst a practical subject clearly warrants applied research, as soon as more basic problems are encountered they automatically fall within another scientific discipline. This can only worry those who believe that subjects are water-tight compartments with no overlaps.

Kinds of Research

Opinions differ greatly as to the necessity of classifying research into different categories and, having done so, as to what they should be called. The idea that there is only "good" and "bad" research illustrates the impatience with which some view the whole discussion, but, in fact, also suggests that research can be unambiguously put into one or other group. The great range of kind and quality in research makes this unlikely and, indeed, it is still necessary to state for what (or for whom) the research is "good" or "bad".

One thing is clear: whatever the merits of classifying research and whether research can be best described in one way or another, there are certainly different reasons for undertaking it. This suggests that it might be better to base research categories on these different reasons, without any implications as to the quality or nature of the research involved in each category, the kind or quantity of the resources required to sustain it or the quality of the people who engage in it.

There are, of course, many reasons why someone engages in a particular piece of research: some of these reasons are quite personal, of the kind that influence people in deciding whether or not to engage in any other activity. The reasons that concern us here are those adduced for undertaking research at all, irrespective of personalities and excluding reasons of commercial or government prestige (the reasons in these cases are for spending resources in that particular way, rather than reasons for the research). These reasons are more general than the objectives of particular pieces of research but both need to be stated very clearly. Although this need is self-evident in applied research, it is surprisingly hard to specify. This is chiefly for two reasons. First, it forces a clarity that may at times be uncomfortable and exposes the fact that there are generally several reasons and objectives and no wish to choose between them. Secondly, it forces a recognition of the high cost of obtaining even very limited amounts of information.

Before proceeding with a discussion of the classification adopted here, it is necessary to emphasize that other classification schemes will be better for other (quite legitimate) purposes and that alternative ways of classifying research for any purpose are worth considering. For example, Gordon (1972) has cogently argued that the distinction between basic and applied should be abandoned, since they relate to use or intent, and replaced with distinctions that relate to "process and organizational imperatives". He favours "urgency", referring to the speed with which the results are needed, and "predictability", referring to the extent that the steps necessary to achieve the research objectives are predeterminable. This approach has led to useful findings, such as that, in general, groups do better than individuals in solving problems with predictable solutions (e.g. solving crossword puzzles), whereas individuals do better than groups in solving problems when the answers were not predeterminable (e.g. creating crossword puzzles) (Hare, 1962; Taylor *et al.*, 1958). The implications of these conclusions are considerable. For example, the idea of inter- or multi-disciplinary teams for research into subjects that overlap several disciplines (O.E.C.D., 1970; Moule, 1972; Charlton, 1973) would need to be modified according to the extent to which problems could be regarded as predeterminable.

This well illustrates that the way in which research activities should be grouped may need to vary with the purpose: for the present purpose, the distinction between applied, basic and fundamental seems useful.

Applied Research

The broad reason for undertaking applied research is to benefit practice. But practice itself may have many objectives and there are innumerable different ways of helping to achieve each one.

In Chapter 1 the purposes of agriculture were discussed in some detail and without an appreciation of this range of purposes it is difficult to think sensibly about research aimed at helping practice to achieve them. It may be that they cannot all be achieved simultaneously and that some of them are incompatible. So it is essential to be sure at the outset of applied agricultural research whether the object is to benefit the producer financially in the long or short term, or the food consumer, or the government in terms of improved balance of payments or independence of imports, or to create employment opportunities or to improve the living standards of the people within a region.

Since economic objectives are often considered to be the most important, it may be worthwhile pursuing this aspect a little further. If it were possible, by research, to put one farmer in a position to make much more profit from one particular enterprise, several processes would immediately begin. One would be that other farmers would adopt the same methods and erode the first individual's competitive position. This might be followed by a reduction in the price of his product and thus of the profitability of his enterprise. If profits remained high, other farmers would start such enterprises and still further spread the total gain, erode competitiveness and increase demand for the resources used in this particular enterprise. The last effect would spread the benefit to yet others but also reduce profitability. Since total production would be increased by most of these changes, the price would ultimately fall and, if it did not, surpluses would result.

Thus *merely* making a greater profit might be of little benefit, except to a minority in the short term. On the other hand, if research results in the more efficient use of resources, including money, the benefit should be lasting and widespread. Results will clearly not be adopted if they lead to loss of profit and will be adopted faster if profit is substantially increased, but it is doubtful whether a marked and sustained increase in individual profits is a realistic objective of a particular piece of research: it is, of course, a normal commercial objective of a continuous research activity.

Increased efficiency in the use of resources, to produce what the community wants and is prepared to pay for, is thus a key objective. But there are many resources and increasing the efficiency with which one of these is used may reduce the efficiency with which others are utilized. Thus, it is no use stating the object of applied agricultural research as, for example, more beef. In no case is it intended that this should mean more beef without regard to the cost: it always has to mean more beef per unit of some resource (including time and money). But it rarely means per unit of one resource only and achieving a statement of research objectives in these terms is not at all easy. Furthermore, it has to be recognized that these objectives cannot be ignored, they have to be stated for all applied research. For example, it can be argued that applied research should concentrate on the current problems of existing systems and that these practical problems can be clearly stated. In a sense they can, but not always in the sense that allows a relevant solution, or, more strictly, excludes irrelevant ones.

Suppose the problem to be a disease that occurs within a well established system of milk production. The practical issue can be stated as how to control or prevent it, but at what cost? These questions can only be answered by reference to the ultimate objectives of the system: it can then be judged what effect any postulated solution would have on these objectives by determining the effects on all the relevant inputs and outputs.

The judgement as to which inputs and outputs are relevant is a matter for those commissioning or initiating the research. Questions cannot usefully be posed without some restriction on the range of inputs and outputs to be examined. Furthermore, the *only* way to take into account *all* inputs and outputs is in financial terms.

Of course, it is possible to develop "multi-goal objective functions that rank other, non-monetary objectives such as environmental quality or income distribution just as explicitly as economic efficiency" (Schramm, 1973), but this has to be done by weighting such objectives in a way similar to assigning monetary values or by imposing them as constraints to the efficiency functions. One solution is to determine, in benefit-cost analysis, the consequences to these other objectives of alternative ways of achieving degrees of economic efficiency. By starting with an unconstrained income-maximization plan and then putting emphasis on one or other of the "societal" objectives, it is possible to measure the income gain that would have to be foregone in order to achieve them.

Thus, economic assessment is one of the indispensable tools of applied research. Since this proposition is so frequently misunderstood, it must be emphasized that it is one tool amongst many, that there is no particular merit in not using it and that what is done with the results is entirely up to the research workers involved. Of course there are dangers if anyone places undue weight on an economic assessment, as there are for any other single assessment judged in isolation. But this should not be allowed to distort the role of so valuable a technique. After all, the economic outcome is of some consequence and sensitivity tests that tell us which factors affect it, and by how much, are of great value.

It is true that economic values change with time, often quite quickly, unexpectedly and by unpredictable amounts; today's costs and prices are therefore argued to be irrelevant. The same is true of climatic variables but they cannot be ignored for these reasons. In both cases, we require the response of outputs to inputs over a range of relevant values for the important variables, including economic ones. Indeed, this is the essence of applied research, to determine the responses of the important outputs to changes in the main inputs, within a relevant context.

This has always been so but traditional experimental approaches have experienced considerable difficulties in dealing with the agricultural situation.

This can best be illustrated by considering the approach of varying one factor at a time and leaving all the others constant. First of all, in biological systems generally and in agriculture especially, the other factors will not stay constant. Thus if we place six calves in a field, whatever we do—even if we do nothing, the quantity of grass will not remain constant and nor will its growth rate. The weather will vary, the calves will probably get larger and the amount of faeces on the field will increase. Even if we hold the *number* of calves constant, the *meaning* of this number changes, since the total weight of calf increases and so does their food requirement. Relative to the food supply, the changing food requirement may differ still more. Now, suppose that we vary the number of calves, in order to study the importance of this factor: quite clearly the meaning of this experimental variation will change with time and the effect over a period of time will be an almost unique cumulative result of all the changes that have taken place over that period.

There are also considerable problems of defining relevance (Morley and Spedding, 1968) in agriculture. This was one of the reasons for arguing (Chapter 2) that there was a need for a systematic way of deriving relevant sub-systems that could sensibly be studied. In fact, the general finding is that relevance and control are difficult to obtain at the same time. Control is wanted in order to avoid the situation where everything is varying simultaneously and the total system is never the same. For many agricultural systems, however, this variation is characteristic and reducing it makes the situation less like the real one. Thus, the effect of temperature on a mountain sheep can be studied most precisely in a controlled experiment in which the temperature is held constant for appreciable periods and such things as wind speed, wetness of the fleece, movement of the animal and food supply are tightly controlled. But the relevant situation is not like this. On the mountain, wind and rain vary rapidly, the sheep moves about and takes shelter, and even lying down may alter its local micro-climate. It would be virtually impossible, however, to measure the effect of temperature in this relevant situation and certainly not with any precision. But if the effect is measured in the controlled situation, is the information obtained of any relevance to a sheep on the mountain?

One point seems clear: if the information is not relevant to the practical situation, the research cannot sensibly be described as applied. The fact that it is carried out with the ultimate aim of benefiting practice does not make it applied research, although it may be argued that the intention was of this kind but the research was poorly carried out. This suggests two propositions. First, that applied research cannot be characterized by the reasons for undertaking it, unless they are stated very precisely. Secondly, "basic" research, under more controlled but less relevant conditions, may have long-term practical benefits in view but its more immediate objectives are the better understanding of a component part or process. This will be further discussed later in this chapter.

Practical situations are not only characteristically variable, they also tend to be "well-buffered" and adaptable. Suppose that we wished to understand how a watch worked. We would not simply take it apart and look at the bits separately. We would have to take it apart, however, in order to see all parts of the structure, but we would try to learn how it worked whilst we were doing so. This requires a kind of simultaneous analysis and synthesis but is, in fact, how we operate when trying to understand how things work. The method does not work, in its simplest physical sense, for living things. Simply pulling a living system apart confers only limited information about probable functions, partly because it involves destruction of the life of the system and prevents us from carrying out the synthesis part of the process.

More important still is the adaptability of living systems. For example, the decomposition of leaf litter in a forest may be undertaken by many different organisms. Let us suppose that there is one dominant species and that we remove it. The other species may then increase their numbers and, given time, the same function may be discharged just as before. By contrast, if we remove an essential part of the watch, it will not work and from this observation we can deduce that its role is essential. In agriculture, it could be argued that, if the system adapted itself so successfully that removal of one component did not alter the functioning of the system, then that component was not essential either. This raises the question of what "essential" means in such systems.

If we removed one leg from a four-legged table and it remained standing, we would not argue that the leg was not essential, for two main reasons. First, we are conscious that we could demonstrate, successively, that none of the legs was essential, but only meaning, of course, that any one of them could be dispensed with at one time. Secondly, three legs are insufficient in many of the circumstances for which tables are used. So the answer is only that in given circumstances three legs are enough: in some special circumstances, two legs are adequate. This is another way of stating the original notion, that changes in outputs have to be related to changes in inputs over a range of relevant contexts. What is needed, in general, is not the relationship between one output and one input in one context, however, and, in any event, there are innumerable such combinations. Generally, the simplest requirement will be a response surface (see Fig. 10.1) but even this will often prove inadequate.

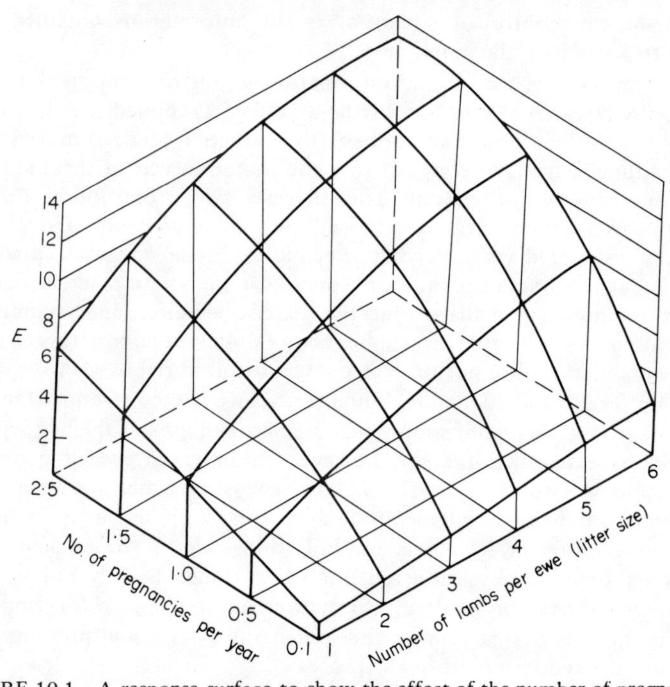

FIGURE 10.1 A response surface to show the effect of the number of pregnancies per year on the efficiency of food conversion (E) by ewes and lambs.

$$E = \frac{\text{carcass output (kg/year)}}{\text{food consumed (kg D.O.M./year)}} \times 100$$

Calculated for Scottish halfbred ewes, all (Suffolk x) lambs being artificially reared. (After Spedding, 1970.)

All such pictures can be legitimately called models but what is now chiefly required is to extend the scope of these models to include all the essential relationships. As discussed in Chapters 2 and 3, only the essential relationships can be handled and the only way to determine whether something is essential or not is by reference to the purposes of the model; these must therefore be very precisely stated.

Thus, the chief conclusion is that the main object of applied research in agriculture involves the construction of models of the major agricultural systems and their constituent sub-systems and processes. The construction of models requires many data, in an appropriate form, from many sources and ensuring the supply of this information can have a considerable impact on research programmes and organizations. There is therefore a heavy responsibility on the model builders in applied research to make sure that their models are the most useful or important of all those that could be developed at that time.

One function of basic research is to supply data to applied workers and model construction has the enormous advantages of making it clear, to all concerned, what data are needed, why, and where and how they will make their contribution to more applied research and thus, ultimately, to practice.

Before discussing basic research, it may be useful to deal with some of the most important misunderstandings that develop between research workers and the users of research results, since these occur mainly within the applied field.

Misunderstanding of Research

Misunderstandings arise for many different reasons: some are misunderstandings of the methods employed in research and others relate to the objectives, purposes, priorities and the relevance of research. Practical men do not generally appreciate the sheer cost of obtaining information by research. Many research workers are ignorant of the monetary cost but most know the very large effort that has to be made in order to acquire quite small amounts of new information. Years of work may be needed to further understanding in ways which may simply appear to consist of a slightly clearer statement of a set of relationships or a somewhat more accurate description of a process. This means that research workers can do very little in relation to the total need: they can do very few of the experiments that they would like to do and that could be justified on the grounds that they would provide information that is both lacking and needed.

Much of research planning has therefore to be related to the determination of relative priorities and one of the criteria that could reasonably be adopted for applied research is "practical relevance". Practical men are not necessarily very good at assessing priorities on the basis of general relevance or at stating their own problems in a manner that makes it possible to conduct research directed to solutions. On the other hand, practical men can see perfectly well whether solutions are available and can test possible solutions to see if they work. Many research workers know themselves to be so divorced from practice that they cannot proceed on their own either.

So both sides need each other and a greater interchange of views is desirable, provided that these exchanges are not just confrontations. What is needed is a better understanding of each other's points of view and both the methods and language of science may act as barriers to achieving this. The argument here is not that better understanding is required simply to improve extension work, or even to make research programmes more responsive to practical need. It involves these ideas but also, more generally, that a subject like agriculture, with a strong practical aspect to it, must be developed by all concerned. It cannot simply be an academic matter but requires contributions from the practitioner as well.

Indeed, perhaps the two greatest needs are for effective contributions from the practitioner on the one hand, and from the theoretician on the other. Curiously enough, the practical man is somewhat anti-theory and thinks of it as opposed to practice. Now, good theory ought to be borne out by practice; it is untested hypothesis and speculation that may conflict with it. Farmers, like others, speculate freely about their problems but tend to be suspicious of theory.

However, as pointed out in Chapter 5, research cannot be conducted on the actual farmer's cows, although it is hoped that the results will apply to them. Similarly, although the research has to be carried out on *somebody's* actual cows, it is certainly hoped that the results will not only apply to them. In short, the information required relates to theoretical rather than actual cows and, because it does so, it should apply sufficiently well to both sets of cows and to many more of the same kind. Theory is a powerful tool and needs recognition: subjects advance by development of theory, rather than by the accumulation of lore relating to particular experiences.

The language of theory, however, may be more difficult for practical men, partly because it is more abstract. Thus cow A and cow B may be the subjects of statements that should apply to all sufficiently similar cows, whilst the farmer might prefer to relate the statement to Daisy, an actual cow. Actually, this example is out of date, simply because farmers have changed. But it would be difficult for a farmer to share the excitement of a biologist who thought he could say something useful about large mammalian herbivores, even though a part of the excitement stemmed from the fact that the statement applied to Daisy, cow A, cow B, all other cows of all other breeds and even to buffaloes.

So part of the language problem is in its abstract nature and part of it is the jargon that creeps into all specialisms, including farming. Technical language is a necessary shorthand and jargon may be defined as unnecessary technical or special language. The fact that hogs in the U.S.A. are pigs in the U.K. and that hoggs in the U.K. are sheep in the U.S.A. is confusing but not a special problem of agriculture. The fact that sheep of the same kind have several different special names within the U.K. is an historical left-over but confuses scientists considerably. More serious is the unnecessarily contorted phraseology of some scientists that impedes understanding between scientist and farmer, between scientists and, often, within one scientist.

Even the simplest language does not necessarily solve the misunderstandings that are concerned with the methods of science. However, as the earlier discussion should have demonstrated, it may well be that agricultural scientists have become over-preoccupied with inappropriate methods, designed for other subjects and inadequate for dealing with agriculture.

There is considerable confusion about the underlying philosophy of scientific method anyway. It may be that some scientists have been over-concerned with methods based on trying to disprove hypotheses and that too many non-scientists have been assuming that the object of applied research is to prove that some proposition is correct. The neutral position, that a scientist is simply trying to find out the truth, is difficult to sustain in applied research since he is also going to argue that the question posed is relevant and has been given priority over others on this basis. It then seems obvious that he probably knows what he would do with the answer and that he probably hopes that it will come out one way rather than another. Furthermore, whenever the stage has been reached at which practical recommendations appear possible, they do have to be tested and the research worker hopes for the expected result: indeed his only real doubt may be as to the precise quantitative expression of the result in a particular year.

It is perhaps not surprising, therefore, that the practical man finds greater difficulty in understanding the objectives of experiments that are primarily designed to establish relationships, and in interpreting their results. This is particularly so where the process of popularizing research errs by emphasizing the more spectacular or bizarre features.

One example will suffice to illustrate this.

In sheep production where meat is the major output, the efficiency of food conversion is important, because food costs represent about half of the total costs, and is greatly influenced by prolificacy (i.e. by how many lambs each ewe produces each year).

A sensible question to pose, therefore, concerns how many lambs a ewe should have per year and, as part of this question, how many lambs a ewe should have in one litter (since sheep are capable of breeding up to twice a year).

Now, clearly this is not a simple question and the answer will vary with the size and breed of ewe, the way it is to be fed, the environment within which it is to be kept, the milk yield potential of the ewe, the weight at which progeny are to be weaned or sold, their growth rate, and on many other considerations.

There are more than 230 breeds of sheep in the world (Mason, 1951) and innumerable ways of keeping them. To answer such questions for particular circumstances is possible in a few cases but a systematic approach is clearly impracticable. This is another example of why it is important to try and understand the principles underlying sheep production, in order to serve a wider range of practice. It is necessary, therefore, to try and distinguish those features of a sheep type that really matter in the context of the question posed. For example, ewes that differ markedly in milk yield will almost certainly differ in their optimum number of lambs for a suckled situation. In economic terms, large ewes cost more to feed so they may have a larger optimum number of lambs than small ewes, but this may depend on many of the other factors as well.

Suppose, however, that we can usefully narrow the range of factors down until an experiment can be performed: it will probably take the form of establishing a relationship between efficiency of food conversion (E) and reproductive rate (the number of lambs per ewe per year). If this is done for sufficient ewes over a sufficient range of litter sizes, a curve of the kind shown in Fig. 10.2 results.

Clearly, the range of litter sizes covered must include 0, 1, 2 and 3, since these are found commonly in practice, but the possibility that the optimum value may be higher than this in some circumstances also needs to be taken into account. Furthermore, if it is possible to obtain litters of six or more, they may be so valuable as to make it worth changing the circumstances or devising new ones. None of this can be known in advance and the point about a curve that includes the optimum is that it also tells you what not to do on either side of it, and the consequences of given departures from the optimum.

But the larger litters may be difficult to obtain and require special hormone

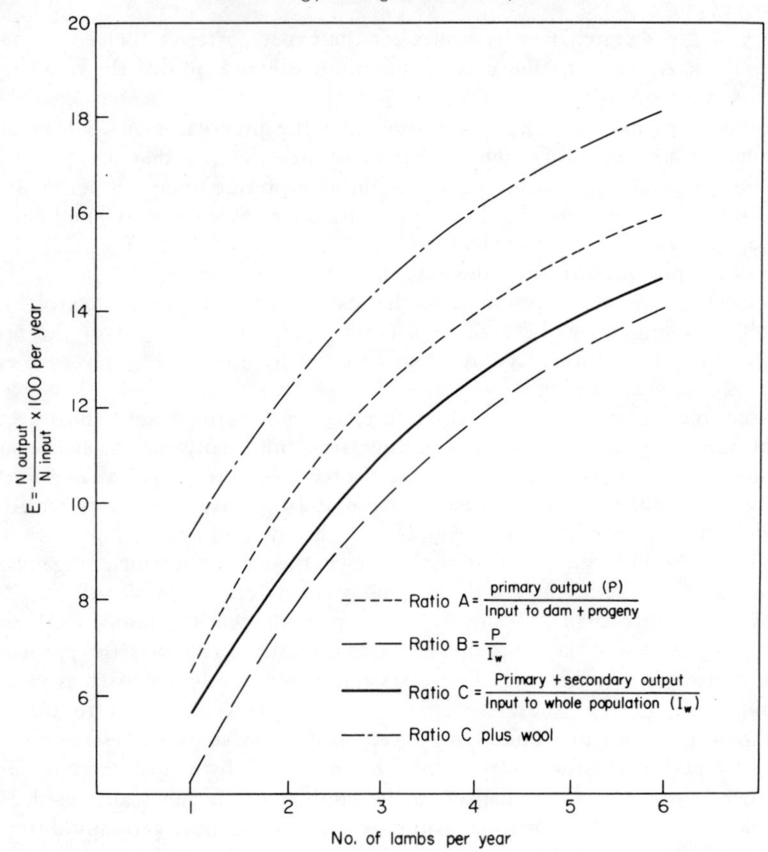

FIGURE 10.2 The effect of reproductive rate on the efficiency of protein production in sheep (from Large, 1973).

treatments: there is no point in breeding a ewe that can produce them unaided if they are not wanted anyway, and it is precisely this that the experiment is designed to determine. Unfortunately, the special problems of larger litters and their more spectacular aspects always lead to greater emphasis on them and greater attention being paid to them. Sextuplets make headlines and singles do not. All this helps to create an impression that the research worker believes that sextuplets are better, that all sheep should have them and that the object of his experiment is to demonstrate this. Furthermore, the scientist himself will generally find the unusual more interesting. In fact, it is the capacity to detect what is unusual that allows his curiosity to consider the implications and possibilities inherent in such departures from the normal.

This (actual) example has been spelt out in some detail because it illustrates so clearly the network of misunderstandings that often cloud the relationship between research and practice and which impede communication. The sad thing is that a research worker who very much wishes to benefit practice may sometimes be the one most likely to be misunderstood, simply because he has an interest in striving to see how his findings can be applied and this may lead him to concentrate on relatively unimportant areas of agriculture, because that is where his results apply.

The danger of misunderstanding is much less with more basic research.

Basic Research

It is not intended to suggest that these research categories can or should be sharply distinguished from each other. Starting with any practical problem, it is possible to ask questions directly concerned with it and then, successively, to ask questions relating to levels of understanding far removed from the original problem. In many research activities, including agricultural research, it is highly desirable that much of the basic work is related, by clearly defined steps, to the applied research from which it arose and through which it will itself benefit practice. Even when all these steps and connections are clearly established, however, there comes a point at which it is unhelpful to describe the research as applied.

In addition, research is carried out in order to understand the ways in which agricultural animals and plants function and to explore the role of components and processes. Such information is often described as "basic" to our understanding of applied problems and provides essential support for activity at a more practical level.

The danger is that since understanding seems so obviously desirable, it may be thought that it will automatically benefit the applied effort. It may do so eventually, of course, but there can be no guarantee of this. After all, you, the reader, are trying to understand me at this moment, but from a particular point of view and for a specified purpose. You neither wish nor need to understand me as a whole person, if, indeed, this were possible, nor even as, for example, a gardener. It would distort my present purpose if I tried to convey a wider or additional understanding at the same time and it would interfere with your reception if you tried simultaneously to enlarge the kind of understanding that you achieve.

The fact is that all understanding must have a purpose. This is not surprising: since all understanding can be stated in model form and it has already been concluded that multi-purpose models have exactly the same drawbacks. It should perhaps be emphasized at this point, that the number of purposes that a model can serve depends a good deal on the way they are phrased. In some senses, all models can serve several purposes, but there is a limit and no model can serve all the possible, or even probable, purposes for which a system may wish to be understood. "Total" models are never achieved and would be totally unusable if they were.

Thus, basic research is simply thought of as characteristically not directly related to practice but as aiming at achieving an understanding of, for example, biological processes from an agricultural point of view. The definition of inputs and outputs is therefore based on relevance to agriculture and, further, to the particular agricultural purposes selected. To be practically effective, it seems likely that basic research ought to be related to, or at least aware of, the nature of the agricultural system (or sub-system) it is intended to understand.

All this assumes, however, that a planned approach to research priorities is necessarily better. It is clearly necessary in part of the total research effort but it always remains possible that unplanned and even accidental findings may be of greater significance than all the planned approaches. It is important to be clear about precisely which "planning" is meant here. Most research requires careful and detailed planning: the issue here is whether we are clever or informed enough to guide research on to the most productive lines by prejudging their relevance to future agricultural need. It seems likely that observations, even accidental ones, on cows are more likely to result in improvements in cow milk production than are observation on ducks; but is this true if we substitute isolated mammary tissue or bacteria for ducks? It is difficult to be sure and all kinds of unforseeable consequences may flow from any research activity.

In general, if research is not obviously relevant it will have to dig relatively deeply in order to influence applied fields, since it will probably do so by changing understanding at a fairly fundamental level.

Fundamental Research

At one end of the range of basic research the purpose is chiefly in support of applied research and there is no sharp separation. At the other end, the role is similar to that of fundamental research and, again, no sharp boundary lines can be drawn (OECD, 1972).

Fundamental research implies, to most people, very abstruse thinking about very detailed levels of understanding. It is more commonly visualized as ultra-mathematical or ultra-microscopic than dealing with, for example, buffaloes. But, of course, applied, basic and fundamental are relative terms and what is a very detailed and fundamental issue to a farmer or a scientist concerned with whole cows may be relatively applied to a cytologist or a biochemist.

Thus, one of the greatest weaknesses of these terms is that they mean quite different things in different subjects. However, this could be a strength if it firmly demonstrated that you cannot label *people* in this way, because the same man doing the same things could be labelled with each of them in turn, within different subjects. Within subjects like the biology of agricultural systems, applied research is chiefly concerned with the operation of biological systems within agriculture, and basic research includes the biology of component animals, plants and processes. What then is the place, nature and role of fundamental reasearch?

It must be to speculate freely, uninhibited by economic or practical considerations, or even by current biological possibilities (hence it is sometimes called "speculative" research). It harnesses unfettered curiosity and has greater freedom to consider propositions that may appear nonsensical. This is its chief merit, for clearly the most far-reaching propositions must all seem very strange and probably ridiculous when they are first proposed. This is not quite true, in the sense that some simple propositions of great importance seem quite obvious once they are pointed out but may remain hidden for long periods otherwise. This latter notion lies behind Szent-Gyorgyi's description of research as "looking at what everybody else has looked at but seeing what no-one else has seen".

The scientific character of the activity rests on this combination of curiosity, speculation and rigorous testing (usually by experiment) of the resulting hypotheses. Since experiments tend to occur at stages in the speculative process, they may well be designed primarily to check that the speculation so far is not false. If it cannot be disproved, there are grounds for proceeding with further speculation, in whatever direction curiosity leads.

All this having been appreciated, it is usual to say that this is fine but few people are good at it and, in any case, we cannot afford to employ a lot of people and facilities just to satisfy curiosity. This ignores the practical benefit because it is very difficult to be sure that there will be any. Over the whole of such an effort, it would be expected that there will be large potential practical benefits, though these may not be exploited at all, or, at any rate, by those who funded the fundamental research.

So, in fact, we cannot be sure what is the best approach and it seems probable that a balance between them would be most rewarding.

The Indivisibility of Research

Since no sharp distinctions can be drawn between different kinds of research and different approaches are apparently required, it is suggested that research should be considered as a whole. Furthermore, this indivisibility of the research activity within a subject, and the dependence of the nature of the research on the nature of the subject, re-emphasizes the value of considering research, as well as extension (see Keyworth, 1973) and teaching, as an integral part of the subject.

Indeed, one could go so far as to say that a subject that is not the basis of teaching, extension or research does not exist effectively and that the lack of any one of these components greatly impoverishes it.

References

Charlton, P. J. (1973). *N.Z. Jl agric. Sci.* **7** (4), 151-158.

Gordon, G. (1972). *R & D Management* **2** (1), 37-40.

Hare, A. P. (1962). "Handbook of Small Group Research". N.Y. Free Press of Glencoe.

Keyworth, W. G. (1973). *Ann. appl. Biol.* **75**, 155-164.

Koestler, A. and Smythies, J. R. (eds) (1969). "Beyond Reductionism". Hutchinson, London.

Kuhn, T. (1962). "The Structure of Scientific Revolutions". University of Chicago Press, Chicago.

Lakatos, I. and Musgrove, A. (1970). "Criticism and the Growth of Knowledge". Cambridge University Press.

Large, R. V. (1973). *In* "The Biological Efficiency of Protein Production". (J. G. W. Jones, ed.) Cambridge University Press.

Mason, I. L. (1951). "A World Dictionary of Breeds, Types and Varieties of Livestock". Technical Communication No. 8, Commonwealth Agricultural Bureau, Edinburgh.

Morley, F. H. W. and Spedding, C. R. W. (1968). *Herb. Abstr.* **38** (4), 279-287.

Moule, G. R. (1972). *Proc. Aust. Soc. Anim. Prod.* **9**, 436-440.

OECD (1970). "The Management of Agricultural Research", Paris, 1970, p. 14.

OECD (1972). "The Research System", Vol. 1, "France, Germany and the U.K." Paris, p. 32.

Popper, K. (1959). "Logic of Scientific Discovery". Hutchinson, London.

Ravetz, J. R. (1971). "Scientific Knowledge and its Social Problems". Oxford University Press.

Schramm, G. (1973). *J. Environmental Management* **1**, 129-150.

Spedding, C. R. W. (1970). *Proc. 11th int. Grassld Congr. Surfers Paradise 1970.* pp. A126-131.

Taylor, D. W., Berry, P. C. and Block, C. R. (1958). Group participation, brainstorming, and creative thinking. *Admin. Sci. Quarterly* **3**, 23.

Waddington, C. H. (ed.) (1968-70). "Towards a Theoretical Biology", Vols 1-3. Edinburgh University Press.

Appendix

The Methodology of Circular Diagrams

The Construction of a Circular Diagram

The sequence of steps is quite straightforward, as follows:

1. Define the focal point of interest (F) and the context within which the system to be described operates (see Fig. A.1).

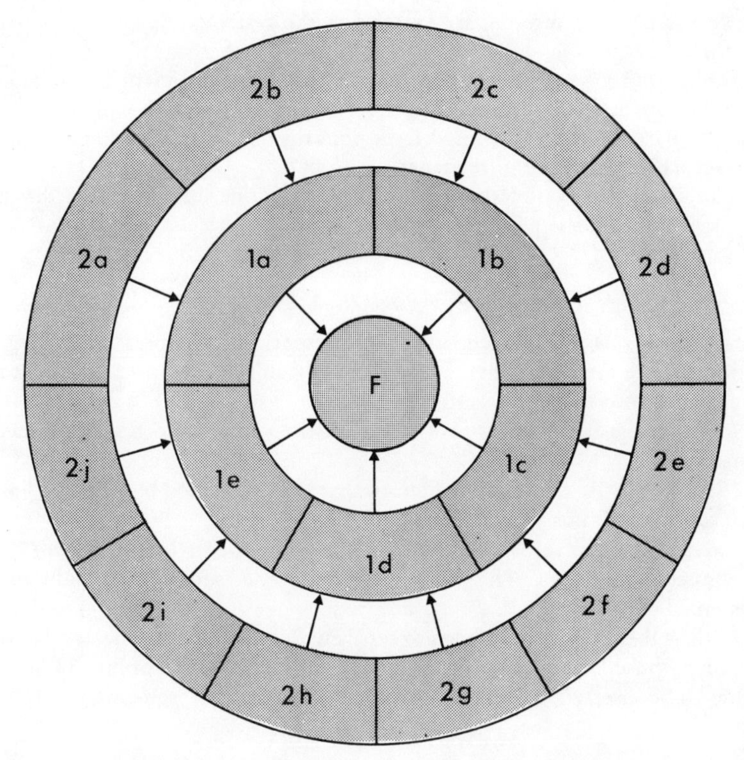

FIGURE A.1 The construction of a circular diagram.

2. Place F at the centre of the diagram, within a small circle. This may be left complete or divided immediately to express a ratio.

3. Subdivide the next outermost circle (1) (that immediately adjacent to the centre) into segments representing those factors (1_a-1_n) that have a direct influence on the centre.

4. Ensure that this circle (1) contains *all* such factors, if necessary by grouping and the use of inclusive terms, such as "other factors".

5. Indicate the influence of each factor (1_a-1_n) on the centre by an appropriate arrow.

6. Indicate influences of any factor on another on the same circle by arrows following circular routes, unless the factors are adjacent, when an arrow at the boundary between them is simpler.

7. Proceed to Circle 2 and follow the same procedure as for Circle 1, except that inward-directed influences should not penetrate to the centre but be directed to items on Circle 1. Any need to go directly from Circle 2 to the centre simply identifies a factor on Circle 2 that should have been included in Circle 1.

8. This process is continued until a natural boundary is met or it is decided to invent one.

In the first case, the necessary condition is that the only factors that influence the last one inserted are external to the system and are thus no part of its essential description.

In the second case, it is decided that the model is sufficiently detailed in a given direction and that values for the last factor inserted will be selected, arbitrarily, at random or at a fixed level, thus displacing any further description of the factors that actually influence it.

9. The diagram is complete when a boundary of one sort or another has been established all round the periphery.

The Extraction of Sub-Systems

A sub-system is considered, for the present purposes, as describing the operation of one factor on the centre, including all relevant interactions. It may therefore be extracted from a system as follows:

1. Identify the factor (f) whose effect on the centre is to be determined or demonstrated.

2. Identify all the component factors that are connected by arrows joining f and the centre, and those connected to such components.

3. Extract the centre, the factor f, all the components identified in (2) and their connecting arrows. This then represents the complete version of the sub-system.

4. If the sub-system extracted is excessively detailed or complicated for some particular purpose, boundaries can be inserted at intermediate points by deciding on values to be inserted (as in item 8 of the "Construction" schedule).

The Definition of Sub-systems

As mentioned in Chapter 2, other workers have used the term "sub-systems" in different ways.

Dent and Bravo (1972), for example, point out that "within the selected boundary for the system and hence the model, there will be ample opportunity to define equally autonomous Systems within the major Systems organization" and conclude that these, "in this context, are best termed sub-systems". Systems synthesis is then visualized as a movement up a hierarchy, a building of systems by the conjoining of sub-systems. Simon (1962) uses sub-systems in a similar fashion.

In the present discussion (Chapter 2), it has been argued that these are component systems and that any component can be viewed in this way.

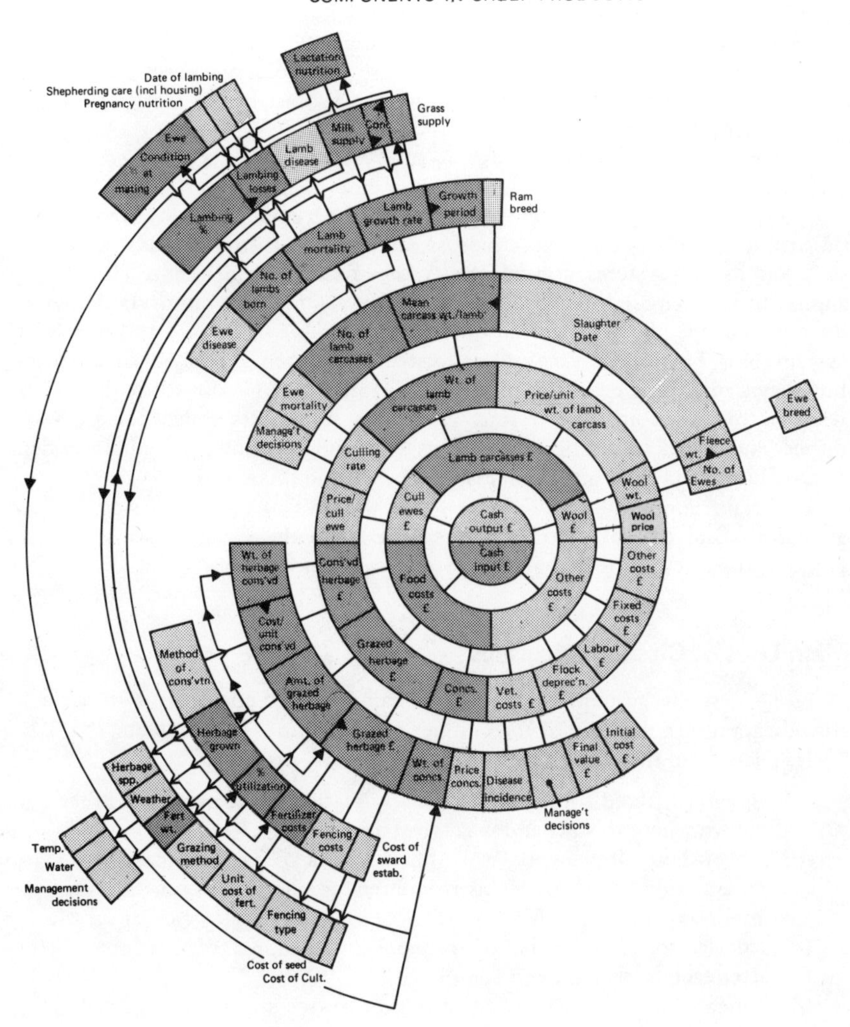

FIGURE A.2 A sub-system (shown darker) of sheep production, primarily concerned with the effect of varying the quantity of nitrogenous fertilizer applied.

However, it is unfortunately true that any of these same component systems may be viewed in many different ways, according to the larger system of which they are a part. Even their essential description may have to vary, since some properties of a system are trivial and others important, *in relation* to the larger system.

I have therefore found it useful to distinguish between sub-systems and components. Miller (1971) also made a distinction between them, as follows: "The totality of all the structures in a system which carry out a particular process is a sub-system" whereas "specific, local, distinguishable structural units are called components or members or parts".

My use of "sub-system" is consistent with this distinction and with the concept of sub-system processes as related to "roles". Thus, a plumbing sub-system has an obvious and definable role within a house and in relation to the main output of a house as a usable system. Similarly, within agricultural systems, the sub-systems extracted by the use of circular diagrams indentify the role of one selected component in the output process of the total system (see Fig. A.2 for a further example.

However, Miller (1971) includes as sub-systems processes that occur in one area of a system and do not necessarily embrace the final output. Thus a "boundary" sub-system is envisaged at the perimeter of a system and a "distributor" sub-system is described as carrying inputs from outside the system or from its sub-systems around the system to each component. This is very similar to the "circulatory" or "nervous" sub-systems of an animal body, which do not end up, physically, at an output, comparable to that of an agricultural system. Nor, however, do natural ecosystems have such a clearly-defined output but they can still be described as systems. It may be that this difficulty can be resolved by describing the output of such systems as simply "functioning", rather like a clock could be described as a "ticking system", quite independent of whether it told the time, even less whether it told it correctly.

The main point to be made here is that the present purpose has to do with agriculture and the sub-systems defined are "agricultural sub-systems". In this context, their definition and method of extraction seem useful.

Main Uses of Circular Diagrams

1. To describe systems that are very complicated and difficult to show in flow-diagram form because of the number of interactions within them.

2. To assist in the description of such systems by providing:

(a) an agreed, fixed focal point, centrally placed on the paper (it is therefore known where to start and what to start with);
(b) a constant framework (of blank circles) that ensure systematic presentation of questions as to content of the system, in order of their importance;
(c) a device for ensuring that interactions are filled in at regular intervals (i.e. after each circle has been completed);
(d) a means of confining the construction to one area (a circle) at a time, so

that the diagram already constructed (inside the circle considered) does not have to be altered;

(e) boundary conditions that determine when the description is complete.

3. To make possible the use of such diagrams to identify, extract and describe sub-systems of particular interest.

4. To serve as a preliminary description of systems or, more probably, sub-systems, prior to the preparation of flow diagrams. It is envisaged that a flow diagram would be necessary before a model was built but that the circular diagram could identify a sub-system to be modelled and could identify its components and relationships. Figures A.3 and A.4 illustrate the way in which a flow diagram may be derived from a circular diagram.

5. It seems likely that models would be constructed as the next stage to circular diagrams: but model building has to be preceded by some method of

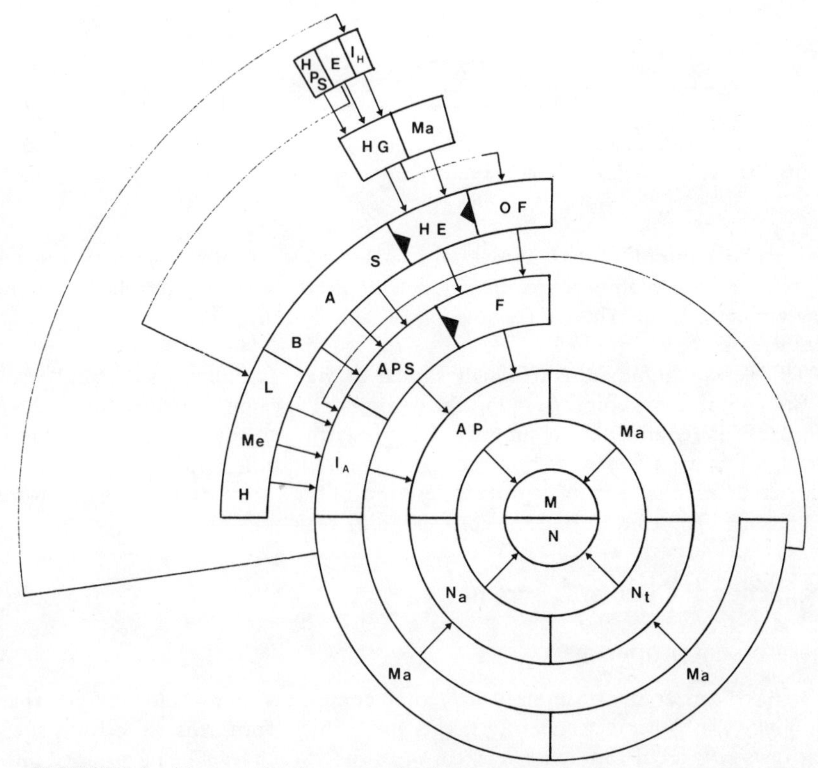

FIGURE A.3 Meat production per unit of nitrogenous fertilizer (per ha/yr). M = meat output; N = nitrogen input; N_a = nitrogen applied at each application; N_t = number of applications; AP = quantity of animal production; APS = attributes of animal production system; I_A = inputs to the animal production system such as Labour (L), Medicines (Me) and Housing (H); M_a = management decisions; F = feed; O.F. = other feeds; HPS = herbage production system with attributes such as Species (S), Age (A) and Breed (B); I_H = inputs to H.P.S.; H.G. = herbage grown; H.E. = herbage eaten; E = environment.

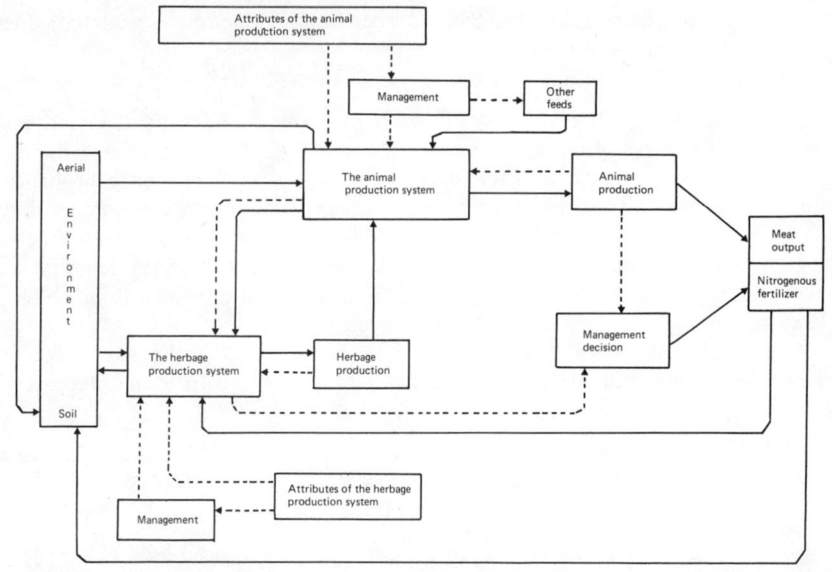

FIGURE A.4 Meat production per unit of nitrogenous fertilizer applied per unit of land per year (derived from Fig. 9.6).

seeing the content of the model in a wider context. Very often this wider context is too complex to be modelled itself but can be described in diagrammatic form. The circular diagram is simply a methodical way of dealing with such complexity.

This general argument is strikingly similar to that of Simon (1962) that "most of the complex structures found in the world are enormously redundant, and we can use this redundancy to simplify their description. But to use it, to achieve the simplification, we must find the right representation". Simon's argument is that we have to substitute a process description for a state description of nature or "given a blueprint, to find the corresponding recipe".

Properties of Circular Diagrams

Hierarchical properties

It has been argued (Simon, 1962) that complex systems and, indeed, that complexity in general, frequently take a hierarchical form; that is, a form such that levels of detail can be recognized, the units at each level being grouped into fewer units at a less detailed level, and subdivided into a larger number of units at a more detailed level. Figure A.5 shows how hierarchies are normally envisaged but many other possibilities exist. Classification charts and family trees are normally hierarchical and they are good examples of relatively simple arrangements.

If we consider a complex system, such as a living organism, it is quite possible

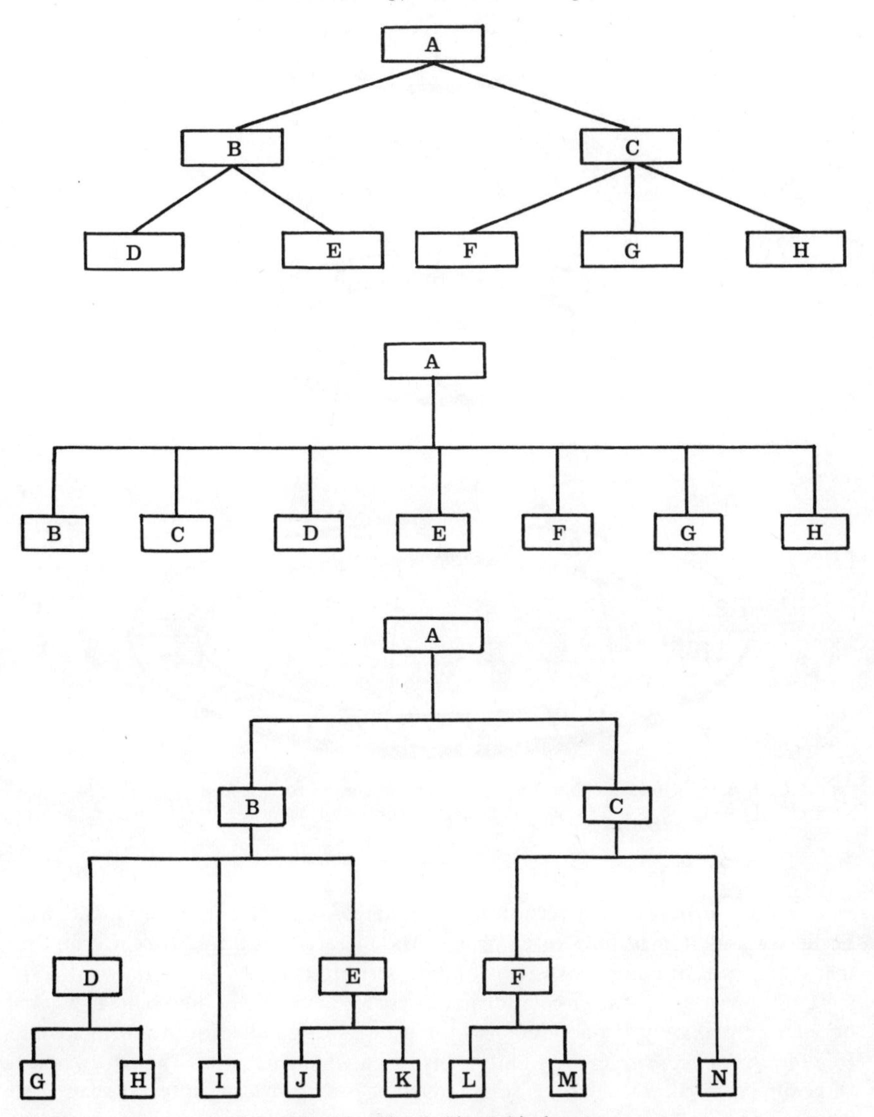

FIGURE A.5 Simple hierarchical arrangements.

to *think* of it in a hierarchical sense—a body composed of organs that are composed of tissues that consist of cells and so on—but there are also a great many interactions that do not follow these simple hierarchical routes and the organism itself may not appear obviously hierarchical. Nevertheless, "matter is distributed throughout space in a strikingly non-uniform fashion" (Simon, 1962): it is, in fact, *arranged* for some purpose.

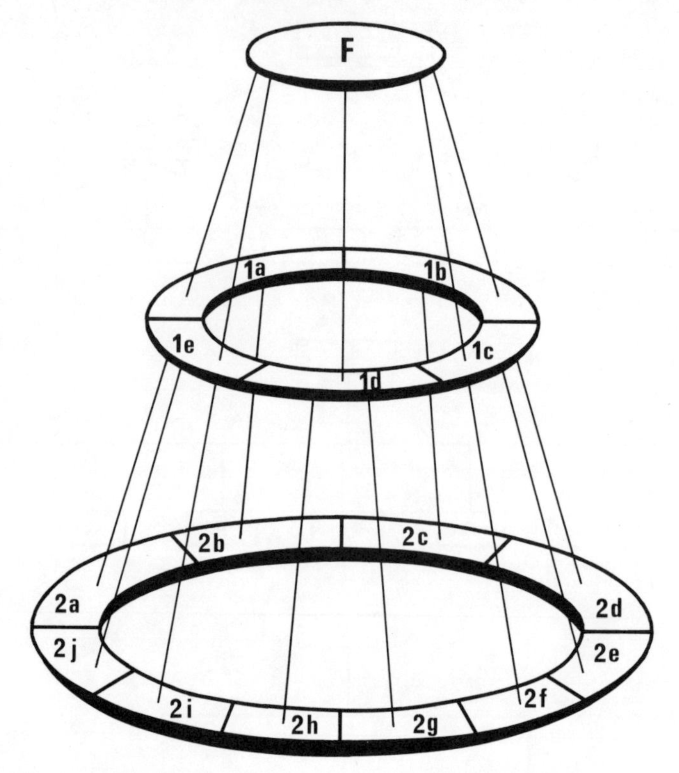

FIGURE A.6 Circles expanded to form a cone, to illustrate their inherently hierarchical form.

For our purposes, of description, for example, a different arrangement may be better and it may help to emphasize the hierarchical framework. Certainly, many diagrams of complex systems find it useful to employ such frameworks. In spite of appearances, the circular diagrams are also essentially hierarchical, as can be seen immediately if one visualizes the centre being pulled out to form a cone (see Fig. A.6). The circles are simply a plan view of such a cone. The latter, being three-dimensional, would have some advantages in providing internal space for inter-connections without encroaching on that used by the main components on the surface.

Similarly, a three-dimensional version could be constructed by retaining the flat circles but taking connecting lines below this plane.

The Nature of Connecting Lines and the Importance of Components

These lines have arrows indicating direction but the lines themselves may relate to material flows or to influences. It might be helpful to adopt the convention, often used in flow diagrams, of using broken lines for influences and continuous lines for material flows.

It has been argued that these connecting lines should never have to penetrate one circle in order to get to another, because they will always act through an intermediate circle component, if the diagram has been correctly drawn. This is not always so, however, for lines connecting components on incomplete circles to complete circles much closer to the centre. This raises the question of relative importance of components and their position in the diagram. It is claimed that the centre is always occupied by the output and that it is helpful to have the centre of interest literally at the centre of the diagram. It would also be helpful if this argument applied to successive circles, but it only does so to a limited extent. It is true that each section of a circle must be less important than the sections immediately between it and the centre (i.e. in the same segment), since these sections include it or represent the way in which it acts on the centre. But the components of any one circle are not necessarily of equal importance and they can be subdivided in several different ways to form the next (outer) circle. The importance of a component (x) on the third circle is not necessarily less, therefore, than one (y) on the second circle, if the second component is not immediately internal to the first (see Fig. A.7).

Since there are alternative ways of arranging the sub-division of components,

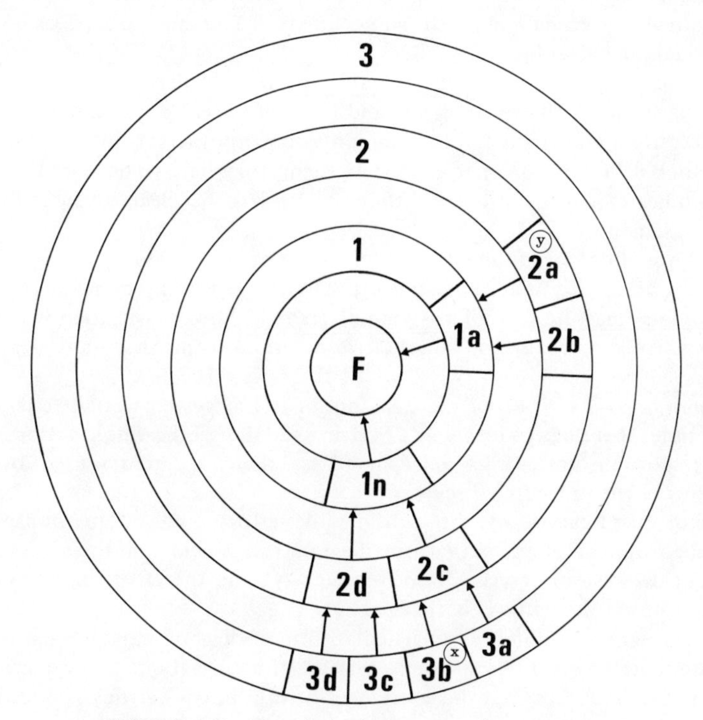

FIGURE A.7 The relative importance of components.

moving outwards from the centre, it would be possible to alter this situation and, if desired, to aim at equal importance for the items on any one circle. In most hierarchical diagrams this involves leaving out some levels and moving directly from a component at one level to a component two or three levels distant. This would leave gaps in the circular diagram but could be overcome by the deliberate use of blanks, if desired.

However, the major difficulty is to assess, at the stage of first construction, the relative importance of components. Frequently this cannot be known; where it is known it is possible to take it into account in several ways, including adjusting the size of the arc represented by a component. Again, the relative size of segments only applies to those on the same circle, since outer circles are larger than inner ones.

The Work Involved in Constructing Diagrams

Anyone who has attempted to describe complex systems in diagrammatic form knows that the amount of effort required is prodigious. It is almost independent of the form of the diagram, since the same number of components and relationships have to be arranged and approximately the same number of decisions taken.

Whether a particular form is found helpful or difficult depends a good deal on the individual concerned and on the subject matter. The same form does not suit all individuals and all subjects equally.

The basic propositions remain the same, however. There is a need to state unambiguously what the components and relationships of a system are, in order to think clearly about them and to communicate with others. If ideas are already clear, it should be a simple matter to state them: they have to be stated in order to communicate to others, even if they are held to be clear to an individual without statement.

It does not altogether follow that all ideas can be stated in diagrammatic form and it may be argued that the mind can handle more complex ideas than can diagrams. This may be true of any one diagram, although even then it may be doubted whether it is true for quantitative descriptions involving any degree of precision.

Furthermore, if, as is often the case, only parts of systems can be studied at any one time, there has to be a way of stating and describing whole systems that makes it possible to extract relevant parts: these are represented by the sub-systems of the preceding discussion.

A major justification of the considerable effort involved in preparing a circular diagram, therefore, is that one diagram can generate all the sub-systems inherent in the system described. So, for each system, the effort only has to be made once, unless the context is changed.

A major reservation about any particular form of diagram must be that it may impose undesirable rigidities on one's way of thinking about a system, in the same way that crutches may be extremely helpful but nevertheless impose on the user a particular mode of progression.

References

Dent, J. B. and Bravo, B. F. (1972). *Agric. Prog.* **47,** 129-138.
Miller, J. G. (1971). *Behavioural Science* **16,** 277-301.
Simon, H. A. (1962). *Proc. Am. phil. Soc.* **106,** (6), 467-482.

Glossary

Agricultural system A system with an agricultural output and containing all the major components.
"Analogue" foods Synthetic foods designed to resemble (and to some extent replace) other, established foods.
Annuals Plants that flower and complete their life cycle in the same year in which they are raised from seed.
Aquaculture The cultivation of aquatic organisms for the production of human food (mainly); e.g. fish farming.
Available visible radiation Wavelengths between 0·4 μm and 0·7 μm also referred to as "usable" radiation: equal to *ca.* 45% of direct solar radiation or *ca.* 50% if the different quality of diffuse component is included.
Biennials Plants that only flower in the year following that in which they are sown.
Biochemical Oxygen Demand (B.O.D.) The amount of oxygen required by micro-organisms, usually in polluted water, slurry or industrial effluent, for oxidation processes. Generally measured as mg of O_2 taken up by 1 litre of the sample when incubated at a standard temperature (20°C) in 5 days.
Biological control The use of one biological agent to control another, generally by predation or parasitism.
Biological system A system consisting essentially of biological processes.
Biomass The total weight of living material, of all forms.
Boundary The conceptual limits of a system, penetrated by outputs and inputs but not by feedback loops.
Browsing A method of feeding by herbivores, in which the leaves and peripheral shoots are removed from trees and shrubs.
Carnivores Animals that feed on other animals or on material of animal origin.
Closed system A system which does not exchange matter with the surroundings; it may, however, exchange energy with the surroundings.
Component An identifiable unit within a system: it may be capable of independent physical existence or be an entirely conceptual entity.
Culls Breeding animals removed from the breeding population, generally on account of some physical or performance deficiency.
Depot fat Fat reserves localized within the animal body and used in times of poor nutrition.
Deterministic A deterministic situation is one in which given inputs lead to predictable outputs.
Detrivores Animals that feed on dead plant or animal material.
D.O.% (or K.O.%) Dressing-out or killing-out percentage, calculated as the weight of carcass per 100 units of live weight.
D.O.M. Digestible Organic Matter – a measure of the nutritional value of food to an animal.
Ecosystems Systems which include both living and non-living substances interacting to produce an exchange of materials between the living and non-living units.

Efficiency A ratio of output (or performance or success) to the input(s) (or costs) involved, over a specified time and in a specified context.

Energy flow The rate of energy transfer between elements of an ecological system.

Entropy The degree of randomness or disorder of a system.

Evapo-transpiration Loss of water by evaporation and transpiration from the above-ground parts of the plant, dominated by meteorological conditions, especially when water supply is not limiting.

Fallow(ing) Resting land from deliberate cropping: not necessarily without cultivation or grazing, but without sowing.

Feedback (loop) The use of information produced at one stage in a series of operations as input at another, usually previous, stage.

Feedlot An area of land, used to accommodate animals (very commonly beef cattle) at a very high density, not contributing at all to the production of animal feed (all of which has therefore to be brought to the animals from outside the feedlot).

First Law of Thermodynamics In all processes, the total energy of the universe remains constant.

Flow diagram The diagrammatic representation, usually with conventional symbols, of the structure of a system in terms of physical and information flows between compartments.

Food, feed A useful but arbitrary distinction is often made between "food", consumed by man, and "feed", consumed by or fed to animals.

Geothermal energy Energy contained in the earth's heat (as steam, water or hot rock).

Grazing A method of feeding by herbivores, characterized by repeated removal of only a part (generally the leaf) of the plant (most commonly herbage, such as grasses and clovers).

Herbivores Animals that feed on plant material.

Hermaphrodite Bisexual. In a flowering plant, having both stamens and carpels in the same flower: in an animal, producing both male and female gametes.

Hibernation Dormancy during winter: metabolism is greatly slowed, and in mammals temperature drops close to that of surroundings.

Hierarchy A structural relationship in which each unit consists of two or more sub-units, the latter being similarly sub-divided.

Homeotherms "Warm-blooded" animals, whose body temperature is maintained above that of usual surroundings.

Information flow Where a component influences another component without the physical transfer of material.

Integrated control Integrated use of both biological and, e.g., chemical methods of controlling pests or weeds.

Leaf area index (L) Total area of leaf per unit area of land.

Maintenance requirement The food required by an animal to maintain its body weight or to prevent catabolism of its own tissues.

Mathematical model A model using algebraic expressions to represent relationships within the system.

Metabolic age Age related to the proportion of mature weight achieved: calculated as age multiplied by mature weight raised to the 0·27th power.

Model A simplified representation of a system (which may be expressed in word, diagrammatic or mathematical terms).

Net assimilation (rate) Mean rate of dry matter production per unit of leaf area.

Nomadism Continual movement, of man and animals, with no fixed settlement, generally in search of food or water.

Omnivores Animals that feed on material of both plant and animal origin.

Open system A system which exchanges matter with the surroundings; it may also exchange energy with the surroundings.

Pair-bonding A reciprocal "mutual attachment" between two heterosexual, sexually mature, organisms such that aggressive tendencies are largely suppressed and sexual ones enhanced.

Perennials Plants that continue their growth from year to year.

Photosynthesis The enzymatic conversion of light energy into chemical energy and use of the latter to form carbohydrates and oxygen from CO_2 and H_2O in green plant cells.

Poikilotherms "Cold-blooded" animals, whose body temperature varies, to a large extent depending on the environment.

Primary production Production by plants: "primary" in being the first use of solar radiation, the main energy source for biological processes.

Productivity A measure of efficiency, relating output of product to the use of a resource (including time).

Relational diagram A diagram used to show the inter-relationships of components and processes in a system.

Respiration The oxidative breakdown and release of energy from fuel molecules by reaction with oxygen in aerobic cells.

Reversible process A process which proceeds with no change in entropy.

Rumen A large additional reservoir in which food is subjected to microbial digestion before passing to the true stomach.

Ruminant animals Those possessing a "rumen" (e.g. cattle and sheep).

Second Law of Thermodynamics The entropy or degree of randomness of the universe tends to increase.

Secondary production Production by herbivorous animals.

Sericulture The keeping of silk moths and their larvae for the production of silk.

Simple-stomached animals Those without major adaptation to components of their diet (especially fibre), such as pigs and man.

Simulation The use of a model, normally mathematical, to mimic or imitate the behaviour of a system which changes through time.

Sink A state variable outside the system boundary, i.e. not quantified, to which outputs may go.

Slurry Agriculturally, usually refers to a mixture of faeces, urine and water, sometimes containing bedding materials (such as straw or sawdust).

Source A state variable outside the system boundary, i.e. not quantified, from which inputs are derived.

Steady-state A non-equilibrium state of an open system in which all forces acting on the system are exactly counter-balanced by opposing forces, in such a manner that all its components are stationary in concentration although matter is flowing through the system.

Stochastic A stochastic situation is one in which a given input leads to a number of possible outputs, each with a probability of occurrence.

Stratification (of sheep breeds) Spatial distribution of the different breeding components involved in sheep production. Objects include the crossing of breeds developed for different attributes and the exploitation of different environments.

Subsystem Often used to describe any part of a system: used in this book in a more restricted sense, to describe a part of a system contributing to the same output as the system itself.

"Support" energy Used in this book to refer to all forms of energy other than direct solar radiation. It thus includes the products of past solar radiation, such as the "fossil" fuels (oil, coal and gas).

System A number of components linked together for some common purpose or function.

Systems analysis The study of how component parts of a system interact and contribute to that system.

Third Law of Thermodynamics The entropy or randomness of a perfect crystal is zero at the absolute zero of temperature.

Transhumance Situation in which farmers with a permanent place of residence send their herds, tended by herdsmen, for long periods of time to distant grazing areas.

Validation The process of assessing the accuracy for a given purpose of a simulation model by comparing the model's predictions with independent results.

Variable A quantity able to assume different numerical values.

Voluntary intake The amount of feed consumed by an animal unrestricted by the quantity available. Maximum voluntary intake is used to describe the maximum quantity (usually per day) that a given animal can consume when neither quantity, quality nor time are limiting.

Yield Quantity harvested: necessarily related to a specified crop or animal, or to an area, and to a period of time. It can also be used in the same way for non-biological processes, but the word "harvested" might not be used ("obtained" would be more usual).

Index

S